SpringerBriefs in Applied Sciences and Technology

SpringerBriefs present concise summaries of cutting-edge research and practical applications across a wide spectrum of fields. Featuring compact volumes of 50–125 pages, the series covers a range of content from professional to academic.

Typical publications can be:

- A timely report of state-of-the art methods
- An introduction to or a manual for the application of mathematical or computer techniques
- A bridge between new research results, as published in journal articles
- A snapshot of a hot or emerging topic
- An in-depth case study
- A presentation of core concepts that students must understand in order to make independent contributions

SpringerBriefs are characterized by fast, global electronic dissemination, standard publishing contracts, standardized manuscript preparation and formatting guidelines, and expedited production schedules.

On the one hand, **SpringerBriefs in Applied Sciences and Technology** are devoted to the publication of fundamentals and applications within the different classical engineering disciplines as well as in interdisciplinary fields that recently emerged between these areas. On the other hand, as the boundary separating fundamental research and applied technology is more and more dissolving, this series is particularly open to trans-disciplinary topics between fundamental science and engineering.

Indexed by EI-Compendex, SCOPUS and Springerlink.

More information about this series at http://www.springer.com/series/8884

Štefánia Olejárová · Juraj Ružbarský
Tibor Krenický

Vibrations in the Production System

Measurement and Analysis with Water Jet Technology

Štefánia Olejárová
Faculty of Manufacturing Technologies
Technical University of Košice
Prešov, Slovakia

Tibor Krenický
Faculty of Manufacturing Technologies
Technical University of Košice
Prešov, Slovakia

Juraj Ružbarský
Faculty of Manufacturing Technologies
Technical University of Košice
Prešov, Slovakia

ISSN 2191-530X   ISSN 2191-5318   (electronic)
SpringerBriefs in Applied Sciences and Technology
ISBN 978-3-030-01736-1   ISBN 978-3-030-01737-8   (eBook)
https://doi.org/10.1007/978-3-030-01737-8

Library of Congress Control Number: 2018957045

This Springer imprint is published by the registered company Springer Nature Switzerland AG
The registered company address is: Gewerbestrasse 11, 6330 Cham, Switzerland

Preface

The submitted monograph "Measurement and analysis of vibrations in the production system with water jet technology" originated as an output of a project KEGA 006TUKE-4/2017.

The monograph was written at research and experimental workplace in a laboratory of liquid jet, Institute of Physics, Faculty of Mining and Geology, University of Mining and Metallurgy—Technical University of Ostrava. It is intended for the experts in the respective field, for broad professional public and for people concerned over getting acquainted with principles and procedures related to the measurement and assessment of magnitude of acceleration amplitude of vibrations in machining by the application of water jet technology.

The monograph consists of nine separate chapters containing basic information on domain of the application of water jet, sieve analysis and its working principle, measurement and assessment of magnitude of amplitude of acceleration of technological head vibrations of water jet.

The monograph presents actual theoretical knowledge and results of the relevant researches.

Prešov, Slovakia

Štefánia Olejárová
Juraj Ružbarský
Tibor Krenický

Contents

Introduction

The constant development of production machines and of methods of material machining requires also the development of the means of regulation and of appropriate diagnostics. In the machining process, the undesired vibrations occur, which cause deterioration of the quality of products. Furthermore, these mechanical vibrations represent a reason for shortening the service life of equipment and of its parts, threat to safety of operating personnel and of environment or to economic efficiency and to competitiveness.

Knowing the issue of mechanical oscillation, of data collection and of signal analysis represents one of the possibilities of getting acquainted with the principles of oscillation processes in the systems which allows further determination of the means of their elimination. Mechanical oscillation acting upon the equipment must not exceed defined limit. The limits are defined by a standard (for instance, ISO 10816, ISO 13373-1:2002 standard) or experimentally. Exceeding the specified limits negatively influences service life, reliability of structural elements and noisiness. The phenomenon can be eliminated once the spot generating excessive oscillation is localized. Time development, frequency spectra along with the frequency spectra envelopes and Fourier transform are applied in case of localization and analysis. A number of precautions exist which allow the prevention of failure rate of machines by means of the elimination of detrimental effects and assure the increase in quality of products. Owing to constantly improving methods of monitoring and of diagnostics of machines, a correct setting of input parameters can result in achievement of its optimal technical condition. Apart from the increase in quality of products, the precaution shall also assure longer service life of the machine as well as higher degree of reliability. Thus, the maintenance and repair costs can be reduced, and high degree of economic efficiency shall be assured.

As to topic the monograph is divided into nine parts. The first part deals with the issue of water jet technology with description of factors which influence the extent of mechanical oscillation and types of used jets. The second part is devoted to the analysis of the particle size. This part includes the topics such as grain composition of material or classification of methods of analysis of grain size. The third chapter contains basic information on grain size analysis and on its working principle, on procedure of its use, on methods of sieving and on types of analytical sieves. Moreover, it describes basic conditions of sieving. Further chapter contains basic information on performance of measurement such as conditions and spot of measurement or plan of the individual measurement. A detailed assessment of measurement is presented in the sixth and in the seventh chapter of the monograph. The last part revolves around the utilization and assessment of knowledge acquired on the basis of the measurement.

Chapter 1
Introduction into the Issue of Water Jet Machining

The ways of searching for methods of increase of efficiency of the process of machining by water jet, of shortening the time inevitable for cutting the individual components with preservation of the quality of the machined area or of reduction of vibration size have not been completed yet. Vice versa, the technology—besides the unlimited assortment of sectile material—disposes a tool without negative effect upon a workpiece or environment and has become a progressive technology in the course of the last years. The infinite or progressive technology of machining by abrasive water jet has gained an importance by adding the abrasive substances. High speed of outflowing water jet with admixture of abrasive allows intensive, economic and ecological machining of almost all industrial materials.

1.1 Characteristics of Abrasive Water Jet Technology

The main areas of origin, formation and effect of high-pressure water jet upon machining material are shown in Fig. 1.1.

Cutting Liquid

Selection of type of working liquid represents a basic factor influencing efficiency of machining technology. Working liquid of abrasive water jet must meet the following requirements:

- low viscosity assuring negligible loses of performance of liquid flow when passing through the piping,
- minimal aggression with respect to metal parts of the equipment, negligible oxidation,
- ability to conform with hydrodynamic characteristics of high-speed jet of small diameter,
- common availability and low price.

Š. Olejárová et al., *Vibrations in the Production System*, SpringerBriefs in Applied Sciences and Technology, https://doi.org/10.1007/978-3-030-01737-8_1

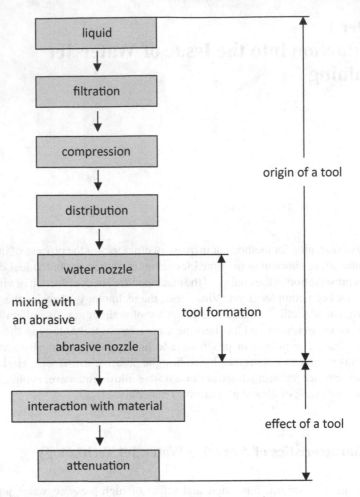

Fig. 1.1 Scheme of origin, formation and effect of abrasive water jet

Basic requirements, which express efficiency of the used liquid, include also ability to preserve compact and linear shape of the jet after being discharged from the nozzle for the longest possible distance.

Filtration and Treatment of Liquid

A long-term run of the equipment designed for machining by the water jet presupposes correct filtration of inflowing liquid. Non-treated water used for drinking or for industrial purposes proved to be unsuitable for a long-term work with pressure of 350 MPa and at speed of 700 m/s. The first equipment used 15–5 μ filters. Apart from the aforementioned water, softening or deionization is performed at times.

Compression—Water Compression

Effect of the water jet upon the machining material is conditioned by intensity of liquid pressure which acts upon particular area. In case of water jet, the pressure of 350–420 MPa acts upon the area of 1 mm². The task of a compression system is to compress the liquid so that a constant effective value of pressure is reached. For that purpose, a piston-type or a multiplier pump is used. The action of a hydraulic multiplier is based on balance of forces acting upon unidentical areas of a differential piston. It reaches the pressure of 300–700 MPa. Figure 1.2 shows a scheme of the hydraulic unit with the multiplier. The following is applicable for water pressure P_2:

$$P_2 = \frac{S_1 - S_2}{S_2} * P_1 \tag{1.1}$$

with

P_1—pressure acting upon the piston with the area of S_1 and
P_2—pressure acting upon piston with the area of S_2

The amount of liquid fed by the multiplier is not continual. Therefore, to balance the pressure fluctuation and to assure balanced feed of liquid, it is included into a circuit by a mechanism. Constant pressure is controlled by an output manometer.

Distribution of Compressed Liquid

Compressed liquid is fed through a high-pressure piping to a movable device of a cutting head. The high-pressure piping is made of steel AlSi 309 = STN 17348.1, 17251.1 standards.

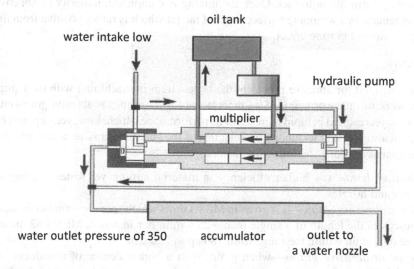

Fig. 1.2 Scheme of equipment designed for water compression with the double acting multiplier

Water Nozzle

A water nozzle is located at the inlet to the cutting head in which the transformation of static into dynamic energy of the jet can be observed. The water jet directs focusing of the jet in order to generate compact jet of liquid. Generation of the continual homogeneous jet represents basis of a machining tool shape which is consequently copied into cutting material. The water jet is an essential component for water jet machining used with both equipment types—in case of pure water machining and in case of abrasive water jet machining. The water nozzles for pressures of up to 150 MPa are made of stainless steel, and nozzles for pressures of up to 250 MPa are made of cemented carbide or ceramics. The nozzles designed for higher pressures are made of sapphire, ruby or synthetic diamond. Service life of the nozzle is conditioned by nozzle material, by used pressure and by quality of the applied liquid. Change of the inner diameter and of shape of the nozzle influences the cutting performance, the quality and the shape of a cut, roughness and perpendicular of the cutting area and the reached cutting depth and size of vibrations.

Mixing of the Water Jet with the Abrasive

The process of mixing of the abrasive material—of a grain and of pure water jet—is performed in a mixing chamber which is located in the cutting head. When the liquid flows towards widening space of the chamber, the lamellar flow changes into a turbulent one, partial swirl of the liquid can be observed, and negative pressure occurs in the chamber. The negative pressure draws the respective abrasive in through the piping which is then mixed with the water jet.

The water jet delivers part of its kinetic energy to abrasive particles, and thus, they jointly proceed to a rectifier tube which is also referred to as an abrasive nozzle; there they form the liquid jet. Once the mixing is completed, majority of abrasive grains remain in a wrapping surface layer of the jet which is rather positive from the point of view of cutting effect.

Abrasive

Costs intended for abrasive represent the largest item in machining with the equipment working on the principle of water jet technology. Adding the abrasive grains into the water jet resulted in rapid increase of its performance which, however, depends on the used abrasive and its quality. In selection of the suitable abrasive, a compromise must be made among the following factors:

- *abrasive hardness*—higher efficiency in material cutting yet faster wearing of tubes and nozzles,
- *size of abrasive grains*—it is given in MESH units which specifies number of sieve meshes in the length of a single inch −25.4 mm. For instance, MESH 82 means size of a grain within the range from 180 up to 220 μm,
- *shape of abrasive grains*—when grains with a higher degree of roundness are used, better surface roughness is achieved at the expense of the reached depth of the dividing cut,

– *abrasive mass flow*—is the amount of added abrasive per specified time unit,
– *environmental friendliness and price of the abrasive.*

Abrasive Nozzle

Basic requirement for abrasive nozzles in which the process of a final jet shape formation occurs is a high degree of abrasion resistance.

Therefore, the abrasive nozzles are produced in powder metallurgy from wolfram carbide, from boron nitride and from a film created in an ion implementation or also from cutting ceramics. Service life of the abrasive nozzle in operation is expressed by a number of working hours of machining with a slight change of its inner geometrical shape (for instance, change of diameter by 0.05 mm) which can be then observed in a change of the outflowing water jet and consequently in a change of shape and size of the formed cutting gap.

The following flows out of the abrasive water jet with production diameter of 1 mm (pressure of 350 MPa, Garnet abrasive with MESH 80, size of grains reaching 180 μm):

– 10–5 h jet with Ø 0.9 mm,
– 20–30 h jet with Ø 1.0 mm,
– 40–60 h jet with Ø 1.1 mm,
– 10–20 h jet with Ø 1.2 mm,
– 15–20 h jet with Ø 1.4 mm.

Maximal service life of tubes made of composite carbide with a film ranges from 100 to 150 h.

Jet Acting as a Machining Tool

When the abrasive jet has flown out of the abrasive nozzle, its performance can be characterized by the following:

– spreading of the jet in the area between the nozzle and the material,
– interaction with material,
– attenuation of residual kinetic energy.

Interaction with material is a connection of several mechanisms, i.e. of machining, fatigue and refraction. The following influences the participation of the individual mechanisms in final deformation:

– angle of incidence of abrasive particles,
– size and shape of abrasive particles,
– kinetic energy of abrasive particles,
– material properties of abrasive particles,
– material properties of machining material.

| new working grating | used working grating |

Fig. 1.3 Consequences of effect of the abrasive water jet upon working gratings after cutting of the machining material during jet attenuation

Water Jet Attenuation

Water jet attenuation can be observed in the area of a retaining tank filled up to minimal height according to working pressure of the jet. Attenuation is performed also with continuation of the jet with abrasive in connection with a supporting grating as it is shown in Fig. 1.3.

1.2 Influence of Diverse Factors upon Mechanical Oscillation

In the process of cutting and dividing of materials by the water jet technology, a number of factors affect mechanical oscillation. The factors influence the magnitude of amplitude A or frequency f.

The identification of the factors may prevent undesired values of mechanical oscillation, and consequently, the factors can influence in a higher or in a lower degree the quality of machined surface or cutting speed. The following can be ranked among the factors (Fig. 1.4):

– stiffness of technological system,
– stiffness of equipment,
– cutting speed,
– cutting depth and shift,
– machining and cutting material,
– working pressure,
– distance between the nozzle and the machining material.

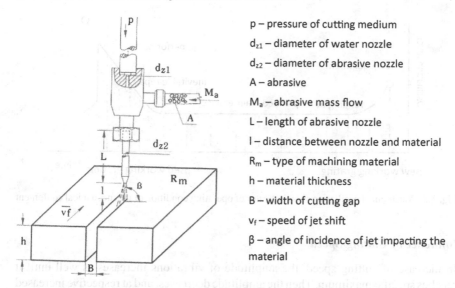

p – pressure of cutting medium

d_{z1} – diameter of water nozzle

d_{z2} – diameter of abrasive nozzle

A – abrasive

M_a – abrasive mass flow

L – length of abrasive nozzle

l – distance between nozzle and material

R_m – type of machining material

h – material thickness

B – width of cutting gap

v_f – speed of jet shift

β – angle of incidence of jet impacting the material

Fig. 1.4   Adjustable parameters in machining of material by abrasive water jet

The effect of the factors leads to formation of a dividing plane as wrapping area of trajectory of the abrasive jet movement. The working jet consists of water (74%), abrasive (23%) and air (3%).

Influence of Stiffness of Technological System

The main method of reduction of mechanical oscillation of the entire system or of its parts rests in increasing of stiffness of the system. In case of changes of material stiffness of the workpiece, the change of amplitude and of frequency of the entire system can be observed along with occurrence of unstable cutting process.

Influence of Equipment Stiffness

The equipment represents the most significant factor as it also causes oscillation of other parts of the system. The overall quality of the workpiece surface depends on stiffness and stability of the equipment operation.

The nozzle is fixed in the technological head, and during equipment operation, oscillation is transferred by the equipment structure to the nozzle which becomes worn under influence of oscillation. Therefore, nowadays the attention is paid to increase the equipment stiffness. If the equipment is not regularly checked and over-hauled hardly repairable defect can occur which may contribute to putting the equipment out of operation.

Figure 1.5 shows the instance of mechanical oscillation as of an indicator of operating condition of the mechanical equipment in the course of time.

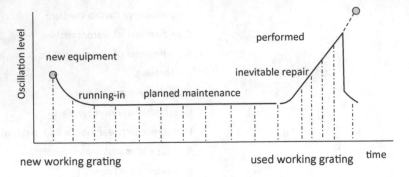

Fig. 1.5 Mechanical oscillation as an indicator of operating condition of the mechanical equipment

Influence of Cutting Speed

In increase of cutting speed, the amplitude of vibrations increases as well until it reaches specific maximum. Then the amplitude decreases, and at respective increased speed, it reaches low values. The influence of cutting speed upon oscillation value proves that certain critical area of cutting speed exists in case of which the amplitudes rapidly increase. The area of cutting speed can be observed under diverse cutting conditions, and the character of the dependence of oscillation amplitude upon cutting speed is the same in each of the cases.

Influence of Cutting Depth and Shift

In case of low values of shift, the oscillation occurs mainly as the result of influence of secondary actuating forces. If the shift values are high, the oscillation occurs as the result of influence of primary actuating forces. The effect of shift upon the vibration amplitude is shown in Fig. 1.6.

Influence of Machining and Cutting Material

When taking into consideration a group of physical and mechanical properties of the workpieces, the intensity of oscillation is influenced especially by toughness of

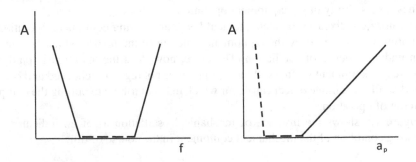

Fig. 1.6 Influence of shift and cutting depth upon the amplitude of self-actuated vibrations

the machining material. The tougher the material is, the worser its machinability becomes. That is caused by its plastic properties, and therefore, the material is more vulnerable to oscillation. Other factor of rather high significance is also material from which the water nozzle is made of.

Influence of Water Jet Pressure

Constantly, increasing water pressure makes the technology progressive. In general, it is applicable that higher the pressure, deeper the cut. In a working process, especially two types of pressure are used which are selected according to the utilization as follows:

- *starting pressure*—serves for material shooting to allow the water jet to cut the hole through the entire material thickness. From this point, the jet moves further on under lower pressure. In case of shooting, the two following methods are employed which are selected especially on the basis of cutting material type:

  (a) stationary method—water jet is focused on a point unless a hole is made through the entire machining material. The method is applied rather with tough material in case of which higher pressure shall not damage the material.

- *working pressure*—water jet rotates around the circle with Ø 1 mm and passes thus through the machining material.

Influence of Working Pressure

In case of higher water pressure, the deeper cut and higher material removal can be observed. The experimental research made by Slaný proved that 380 MPa appeared to be the most efficient pressure value.

Influence of Distance Between Nozzle and Machining Material

In machining by the technology of abrasive water jet, the shortest possible distance from the surface of machining material is recommended with regard to safe movement of the nozzle above the material as threat of nozzle cutting exists. The cutting gaps between the nozzle and material reach the values of up to 5 mm yet ideally of approximately 2 mm. Increase of distance would reduce the process efficiency, and at the same time, it would negatively influence the quality of the cut. And, moreover, the cut would be wider with regard to increasing dispersion of abrasive.

Influence of Shift Speed

Shift speed refers to speed of the nozzle movement towards cutting material. In machining, the water jet stagnates as passing through the material which causes loss of its energy and deflection from its axis by means of which visible grooves occur. The change of shift speed can influence the following parameters:

- cutting depth,
- width and shape of cutting gap,
- surface quality,
- size of vibrations.

1.3   Types of Applied Jets

The following represents three most frequently applied types of jets exist which are installed according to possibilities of classification, structure and requirements:

– system of pulsating jet,
– system of continual jet, and
– system of cavitation jet.

1.3.1   System of Pulsating Water Jet

It employs repeated and short-term lasting of the jet impulse with occurrence of pressure peaks which accelerate widening of the cutting gap. The method appears to be suitable for drilling, cutting and crushing of a stone.

1.3.2   System of Continual Water Jet

It is the most widespread method of cutting of practically all material types. It is typical for constant energy level of the jet in the course of entire cutting process.

1.3.3   System of Cavitation Water Jet

It works upon the principle of local disturbance of material owing to destructive force of cavitation bubbles. It is a case of continual jet with content of cavitation bubbles. Cavitation occurs when the liquid vapours. The air bubbles pass through water which during their extinction induces shocks. Under the influence of the shocks, material gets disturbed.

Chapter 2
Analysis of Particle Size

Grain materials represent the most frequent form of the solid occurrence, and therefor, the grain composition ranks among basic information characterizing the properties of the substances.

2.1 Grain Composition of Material

Grain composition of materials is rather varied. It can differ especially in the following parameters:

- average size of particles,
- size of the most frequented particles,
- range of particle sizes,
- shape and difference between minimal and maximal particle sizes.

Difference between the size of the smallest and of the biggest particles can be minimal yet very often they might be on the level of many orders (Fig. 2.1). The material can contain particles of the identical shape and of approximately identical size or it contains the particles the shape and size of which range within several orders. In general, it can be assumed that oscillations small as to the size are typical for the following:

- synthetic or artificial products,
- products of nature.

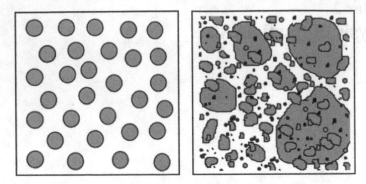

Fig. 2.1 Two extreme cases of composition of the solid

2.2 Classification of Methods of Particle Size Analysis

The analysis of particle size uses several tenths of methods at present. Majority of them can be categorized into three groups—optical, mechanical and gravitational.

The alternative classification of methods of particle size analysis is based on the fact whether the distribution of particle size is achieved through the actual separation of particles or through other principle of determination (Fig. 2.2).

Non-separable techniques can be categorized into two groups as follows:

– analysis of size of the individual particles—pattern analysis,
– analysis of big set of particles—diffraction methods.

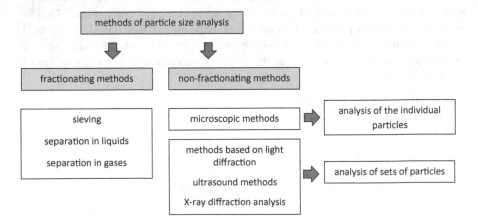

Fig. 2.2 Classification of methods of particle size analysis

The main singularities which make the individual methods different are as follows:

– type of samples which can be analysed,
– extent of applicability defined by minimal and maximal dimensions of particles,
– requirements for sample preparation,
– pace of analysis and of processing of results,
– amount of the sample inevitable for the analysis,
– cost of the instrumentation.

Chapter 3
Sieve Analysis

The sieve analysis represents the oldest, the simplest and the cheapest method of detection of particle size of solids. The method belongs to a narrow group of fractionating or of separating techniques of particle size analysis.

3.1 Principle of Sieve Analysis

The sieve analysis rests in application of a nested column of sieves with the known size of the openings which is elaborated in the direction of gravitation transport of the analysed material into a block with gradually diminishing size of openings (Fig. 3.1). Once the fractioning is completed, certain amount of original sample remains on each sieve which contains particles within limits determined by the size of openings of upper and of lower sieve. The residual amount remaining on the sieve is weighed, and the result is assessed as the weight of fractions with defined range of particle sizes. The result of sieve analysis in the form of weight content of the individual fractions along with gain of actual sample with the specified dimension of particles ranks among the most significant advantages of the method. From this point of view, no other alternative can rival the sieve analysis. The problem rests in impossibility to determine density of material mixture directly in the suspension or air flow. The main disadvantage of sieve analysis of particle sizes is time consumption and destructivity with regard to grain composition of sample.

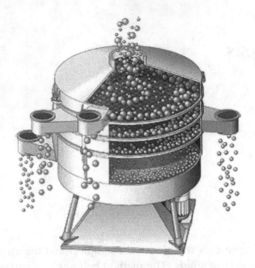

Fig. 3.1 Sieving principle

3.2   Classification of Sieve Analysis Methods

Standard equipment designed for analysis of dry material involves a vibration column with a series of metal sieves. The vibration column includes regulation of frequency and amplitude of vibrations. The analytical sieves are usually encased in circular metal frames, and connection with the adjoining sieves is designed to assure a block of sieves forming a set of stationary segments and to eliminate sample leakage through the area between the individual sieves. One of the sieve analysis devices used in parallel for performance of the experiments is shown in Fig. 3.2.

The sieve methods of particle size determination include a number of known procedures which can be categorized on the basis of the following criteria:

(a) *according to the medium type used in separation of solid*:

  – gaseous (mostly air)
  – liquid (mostly water)

(b) *according to medium characteristics*:

  – stationary
  – movable

(c) *according to state of sieves during analysis*:

  – moving
  – static

Fig. 3.2 RETSCH AS 200

(d) *according to the range of opening sizes in the sieves*:

  – category of standard sieves (meshes with sizes ranging from 100 μm up to 6 cm)
  – category of micro-sieves (meshes with sizes ranging from 3 μm up to 100 μm)

3.2.1 Sieve Analysis Procedure

In sieve analysis, the procedure as described below is followed:

(a) *sample preparation*

The samples are treated by quartering. The sample weight for determination of grain-iness depends on size of maximal grain in material (for instance, up to 2 mm it is 100 g, up to 6 mm it is 200 g). On the other hand, the greater the amount to be analysed, the lower the measurement error is and the more accurate the results are. The wet samples must be dried at first.

(b) *assembly of a sieving set, sieving and weighing*

As a standard, the sieving analysis employs the sieves selected from the following series: 0.063, 0.123, 0.25, 0.5, 1.0, 2.0, 4.0, 8.0, 16.0 up to 40 mm. Naturally, the sieves of other dimensions are produced as well. A set of four or five sieves is inserted into vibration equipment, and a weighed amount of a dry sample is poured onto a top sieve (with the largest openings). The top sieve is selected so that the residues remaining on it equal to zero, and on the lower one the residues should exceed zero. During the warm-off, the suspension is poured onto the top sieve in the course of a run. In case of a dry method, the equipment is put into operation for $10 \div 20$ min, and the period of wet method reaches the value of 10 min. Once the sieving is completed, the individual fractions on the sieves are quantitatively transferred to a tank weighed in advance and there they are weighed. The weights of the individual fractions on the sieves are consequently recorded into a prepared table. In case of the wet method, the oversize ratios are firstly dried and then weighed.

(c) *assessment of sieve analysis*

The sieve analysis is expressed by dependence of weight ratio of the individual fractions on the size of particles. As a standard, it is expressed in two ways as follows:

– *Cumulative*—ratio of all particles in a set of the smaller ones than the given sieve—descriptive ratio is expressed by the relation as follows:

$$y = 100\% - y_r \tag{3.1}$$

with y_r—residues on the sieve

– *relative representation of the individual fractions*

fraction ratio y_i—weight fragment of the i-type fraction. The ratio of weight of fraction m_i falling onto the weight of the entire analysed set of particles m_{vz}

$$\Delta y_i = \frac{y_i}{m_{vz}} * 100\% \tag{3.2}$$

By the HF fractions—ratio of all particles the size of which is bigger contrary to the given sieve and weight m_o refers to the sieve with the size of the openings corresponding with the biggest particle in the sample, i.e. $m_o = 0$ g.

$$y_{ri} = \frac{\sum_{i=0}^{n} m_i}{m_{vz}} * 100\% \tag{3.3}$$

and

$$y_{rn} = \frac{m_0 + m_1 + m_2 + \ldots m_i + \ldots m_n}{m_{vz}} = 1 \tag{3.4}$$

y_{ri} oversize ratio in case of the sieve with size i
y_{rn} weight ratio of all particles which fell through the set of sieves
n number of fractions
m_i weight of the fraction of the i-type
m_n weight of fraction below the finest sieve in a dish
m_{vz} weight of the sample containing all fractions

3.2.2 Procedure in Instrument Granulometric Analysis

A photometric laser method is used for granulometric analysis of a fine-grained fraction. The powder sample is dosed into a dispersant that by means of mixing and ultrasonic disperses the sample in the liquid and pumps it through the cuvette which a ray of light passes through during measurement sequence. Dispersion of the ray of light is assessed as an angle of incidence of the ray of light along with its intensity. The automated statistical processing resulted in a granulometric curve and in a table with fractional ratio of the individual particles expressed in vol. %.

3.3 Sieving Methods

The two following basic types of sieving can be distinguished:

(a) dry sieving (Fig. 3.3)—particles smaller than approximately 50 μm cannot pass easily through the identical meshes as gravitation force is not intensive with regard to friction force acting against the walls of meshes. Except for the aforementioned, the small particles adhere onto the big ones—air jet sieve.
(b) wet sieving (Fig. 3.4)—materials in suspension or powder are used for sieving which tend to aggregate in dry sieving. Those are the particles smaller than 50 μm with considerable electric charge. The electroformed sieves are used which are frequently connected with rinsing, suctioning or vibrations to increase the flow of liquid through the sieves. The sieving is completed when the flowing liquid is pure. The disadvantage is slow sieving—lasting approximately 1–2 h—the individual fractions must be dried prior to weighing.
(c) manual sieving—sample is poured onto a pad. As standard, a binocular magnifying glass or a microscope is used for working. In order to accelerate the work, the method appears to be suitable for a narrow grain fraction. The material is

Fig. 3.3 Instance of dry sieving

Fig. 3.4 Instance of wet sieving

Table 3.1 Sieving period

Size of screen meshes [μm]	Sieving period [min]
40–63	20 up to 30
71–160	10 up to 20
over 160	5 up to 10

removed by a preparation needle the point of which is covered with beeswax (an object is not adherent), by a vacuum needle (an object is sucked due to negative pressure in the needle) or by a single-axis brush (an object is caught due to the existence of electrostatic charge). The method is time consuming yet represents a sole possibility in some cases.

(d) machine sieving—period of sample sieving depends on physical and chemical properties of friable body, graininess and size of screen meshes. Table 3.1 presents the orientation values of sieving period in dependence on size of screen meshes.

3.4 Analytical Sieves

The sieves consist of a hard stainless frame with of high stability. Special attention is paid to screen meshes which is tendered and fits accurately (Fig. 3.5).

The three following types of sieves can be distinguished:

– *Woven* (Fig. 3.6)
  size of meshes: 125 mm–38 μm (± influence of fineness and of material)
– *etched* (Fig. 3.7)
  size of meshes: 500 mm–5 μm (±2 μm)
  shapes of meshes: round, angular, but they dispose of low resistance

Fig. 3.5 Site tissue

Fig. 3.6 Woven sieve

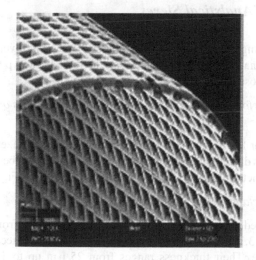

Fig. 3.7 Etched sieve

Fig. 3.8 Perforated sieve

– *perforated* (Fig. 3.8)
  size of meshes: 125–1 mm (± 2 μm)
  shapes of meshes: round, angular, but they dispose of high resistance

   The main advantages of analytical sieves are the following:

– high corrosion resistance,
– simple cleaning owing to high-alloyed stainless materials,
– permanently tight screen texture,
– clear and accurate marking of the sieves,
– maximal stability and optimal tightening when a set of sieves is used.

3.4.1   Types of Analytical Sieves

In general, the cutting of solids includes a number of sieves of diverse structure and sizes. Yet for the analysis of the particle sizes, the three basic types of normalized sieves are used. Those are the following:

(a) *sieves with perforated metal plates designed for the analysis of coarse-grained materials*

   – it is used for fractionation of material with the particle sizes of over 4 mm. Metal plates differ both in a shape of openings, which can be round or square, and in geometry of distribution of openings as shown in Fig. 3.9.

(b) *sieves with rectangular sieve of metallic fibres*

   – they are used for powders with particle sizes ranging from 40 μm up to 4 mm, Fig. 3.10. Metal fibres in the sieves are made of steel, brass or phosphor bronze. Their thickness ranges from 25 μm up to 1.6 mm. General characteristics of analytical sieves are normalized size of openings, tolerated maximal deviations from the size as well as limited number of openings with tolerated deviation. In case of wire sieves, the tolerated deviations of sizes

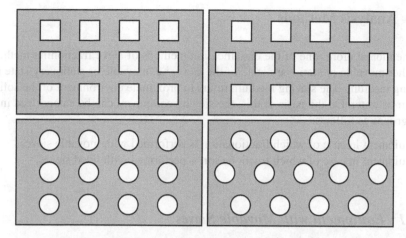

Fig. 3.9 Shape and arrangement of openings in sieves with perforated metal plates

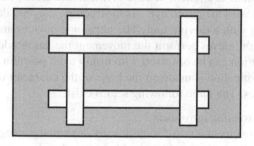

Fig. 3.10 Rectangular method of weaving the normalized wire forces

of openings range from 6 to 30%. The given tolerance increases with the decreasing sized of the openings which relates to the accumulating technical difficulties in production of fine sieves.

(c) *sieves from perforated foils with micro-openings which are designed for the analysis of fine-grained powders*

   – they are used for the analysis of powders with the particle size smaller than 40 μm. It is a case of metal foils in which the openings are made by electro-chemical methods. The size of openings ranges from 3 up to 150 μm so the powders containing the particles with the size ranging from 40 up to 150 μm can be analysed by means of micro- as well as by means of wire sieves. The problem of low mechanical strength of foils used in case of micro-sieves connected with their low thickness is usually solved with a composite double-layer structure in case of which the increase of mechanical strength is assured by the thicker layer with bigger openings. Necessity to improve mechanical properties of micro-sieves is connected with their high resistance in flowing of transport medium.

3.5   Analysis Methods

As mentioned afore, one of the classification methods of sieve fractioning method is either movable or stationary state of the sieve, i.e. movable or stationary state of sieving medium. The sieving medium refers to proximate environment of the solid, i.e. air or water. On the basis of this, the sieving equipment can be categorized into two groups as follows:

- equipment in case of which fractionation is performed with movable sieves,
- equipment in case of which fractionation is performed with fixed sieves.

3.5.1   Equipment with Movable Sieves

It is a case of standard equipment in case of which the sieves perform movement possible to be seen with the naked eye. As to structure, they are designed to assure moving connection with a driving unit. The ultrasonic devices are not included in the group because the sieves perform the movement the amplitude of which is low and thus the oscillation can be observed with firmly fixed position of the sieve.

Further subclassification is made on the basis of the character of movement performed by the sieves. The three following setups exist:

- with continuous rotating movement,
- with vibratory movement—two conventionally determined categories of slow and fast equipment exist. In the first case, it is 1–5 Hz and in the second case, the value ranges from 10 to 50 Hz. Other parameter for classification of the equipment is character of movement of the sieves which includes horizontal and vertical movement as well as the movement with trajectory forming with horizontal plane an angle differing from the one in previous two cases (Figs. 3.11 and 3.12). The last significant feature of the equipment with movable sieves is amplitude of oscillating movement.

3.5.2   Equipment with Fixed Sieves

In case of such equipment, the sieve system is firmly connected with the stable pad to minimize possible movement induced secondarily. The three following groups comprise the category:

(a)   *ultrasonic equipment*—fractionation is realized primarily as a consequence of sieving medium through oscillating movement. The equipment is main representative of the application of methods with the fixed sieves and is applied especially in analysis of fine-grained powders performed with the micro-sieves. The

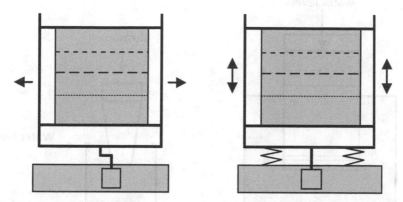

Fig. 3.11 Sieving in consequence of horizontal rotating and vertical vibration movements

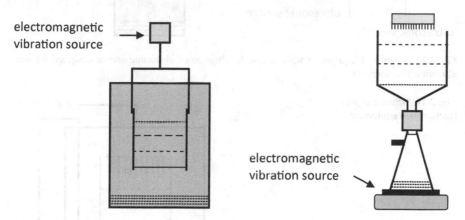

Fig. 3.12 Wet sieving with diverse positioning of electromagnetic vibration source

ultrasound sources are used in case of dry and wet fractionation. The schemes of the most frequently used ultrasonic equipment are shown in Fig. 3.13.

(b) *scheme for wet fractionation* (Fig. 3.14)—fractionation is realized primarily as consequence of movement of the sieving medium by means of unidirectional movement. The method is used only if wet fractionation cannot be applied.

(c) *equipment for fractionation in the air flow* (Fig. 3.15)—fractionation is realized primarily as consequence of sieving medium by means of unidirectional movement. Particle separation occurs due to vertical oscillation movement of fair column.

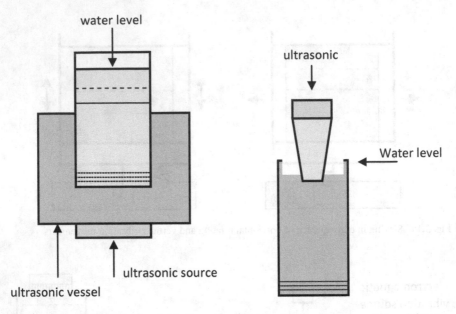

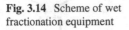

Fig. 3.13 Scheme of equipment with diverse localization of ultrasonic source designed for wet ultrasonic fractionation

Fig. 3.14 Scheme of wet fractionation equipment

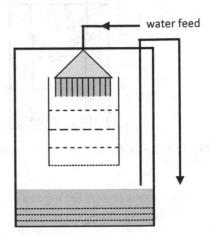

3.6 Conditions of Sieve Analysis

Fractionation of material performed with sieves is based on meeting the three following conditions:

(a) *existence of high probability* that each particle can get to the proximity of sieve openings. The condition is met when the individual particles perform sufficient reciprocal movement.

Fig. 3.15 Scheme of
equipment designed for dry
ultrasonic fractionation

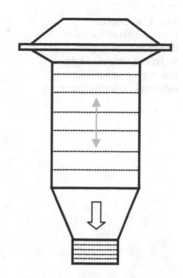

(b) *particles of the size smaller than the size of sieve openings* which they pass
   through. It means that the forces which unite the particles into agglomerates
   are not big enough to prevent gravitational transport of particles through the
   opening.
(c) *particles with the size approximating the size of sieve openings*—the openings
   do not get permanently clogged which means that the probability of occurrence
   of the particles is the same as in case of other particles.

Meeting the aforementioned conditions depends on physical and chemical properties of particles and of ambient medium as well as on intensity and method of
sieving. In general, probability of transport of particles through the sieve depends
especially on the following:

- number of particles,
- size, i.e. distribution of particles,
- shape characteristics of particles,
- intensity of movement of ambient medium,
- overall active area of the sieve and relative area of openings.

The problem of sieve clogging represents a general problem of fractionation of
solids with the sieves yet it becomes considerably exposed especially in case of
fractionation fine-grained material with size of particles below 100 μm. Clogging of
sieves depends not only on the applied sieving method but also on sieve structure and
period of use. The second case is connected with the different influence of transfer
of the primary source movement to the particle movement. The transfer is different
in case of sieves with the solid flexible frame as well as in case of sieves with
the stretched and loose screen. Kinetics of the sieving process in case of clogging
is shown in Fig. 3.16. Deceleration is higher when the size of particles above the
sieve approximates the size of openings in the sieves. A significant role is played by
clogging of sieves which includes gradual reduction of openings and complete loss of

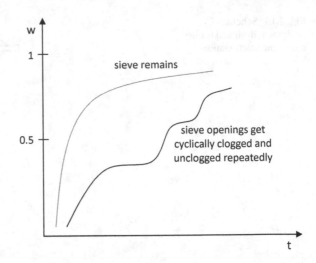

Fig. 3.16 Standard development of curves of dependence of relative weight of sample weight **w** passing through the sieve openings in sieving time *t*

their permeability. The entire process of clogging of pores is the result of "balance" between clogging of pores and their "cleaning "which is typical for a specific type of method as well as for specific type of equipment. The most frequent method used for elimination of clogging of sieves is application of the vibration source moving in direction of particle transport—the movement is usually vertical in case of horizontal positioning of sieves.

In general, the selection of method of analysis of particle sizes as well as of optimal equipment depends on a number of factors. Primary role is played especially by the following character of the fractionated sample:

– grain composition,
– mechanical properties,
– shape of particles,
– their density and reciprocal adhesive forces,
– abrasive properties,
– adsorbing properties.

A number of factors influencing kinetics and result of sieve fractionation represent the main reason of frequently occurring difference in results of grain composition of the same material achieved by means of diverse methods of sieving analysis. One of other significant reasons is connected with considerable differences of diverse methodologies in differentiation of primary particles from agglomerated groups. In case of sieving methods, the difference is clear especially in application of dry and wet methods of fractionation—particles in dry state and particles in suspension are not usually identical objects. Rather often it can be anticipated that even the application of the same method in case of diverse devices shall offer different results. The problem is many times connected with destructive character of sieving methods.

Chapter 4
Vibrodiagnostic Analysis

Any equipment vibrates. Each vibration is specific for the equipment, and vibration is changed in relation to the change of conditions under which the particular equipment operates. The noise produced by the equipment represents only partial information on the state of equipment. Vibration analysis detects a broad scale of defective states of equipment. Increased vibrations may indicate standard wear of components or they can refer to origin of the problem. At the same time, the increased vibrations may signalize a need of inevitable servicing operation. Understanding of origin of vibrations and their development represents a key information for further steps of problem solving.

4.1 Mechanical Oscillation and Its Types

Mechanical oscillation represents movement of a material body in case of which the body is bound with a particular fixed point. The point defines its equilibrium position which means that the material body recedes from and returns back the equilibrium position during movement. The movement of the body or of the system of the bodies which oscillate is usually mathematically expressed by a differential equation or by differential equations. From this point of view, the oscillation can be categorized as follows:

– *linear*—movement of the body is described by a linear differential equation,
– *nonlinear*—movement of the body is described by a nonlinear differential equation,

From the kinematic point of view, the mechanical oscillation is categorized as follows:

– *periodic*—oscillations recur at time and at particular interval:

- harmonic—dependence of deviation of oscillating body on time is described by sinusoid,
- non-harmonic—dependence of deviation of oscillating body on time is described by sinusoid,

– *non-periodic*—oscillations do not recur at time and at particular interval:

From the point of view of damping, the mechanical oscillation is classified as follows:

– *damped*—in oscillation of the body the energy is lost,
– *undamped*—in oscillation of the body the energy is not lost.

According to actuating, the mechanical oscillation is classified as follows:

– *Spontaneous*—oscillation without action of external forces,
– *Induced*—oscillation of the body is induced and maintained by action of actuating forces.

4.2 Basic Oscillation Parameters

In mechanical oscillation, the following basic parameters can be distinguished:

– vibration—periodically repeated part of oscillating movement,
– swing—half of a vibration,
– amplitude—maximal deviation from equilibrium position [mm],
– period—time in the course of which the system performs a single vibration [s],
– frequency—determines a number of vibrations which occur per one second,

$$f = \frac{1}{T} \quad \left[\mathrm{Hz\,s^{-1}} \right] \tag{4.1}$$

– angle frequency $-2\,\pi$ multiple of frequency

$$\omega = 2\pi * f = \frac{2\pi}{T} \quad \left[\mathrm{rad\,s^{-1}} \right] \tag{4.2}$$

Figure 4.1 shows the individual parameters connected with mechanical oscillation.

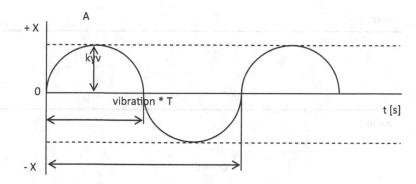

Fig. 4.1 Oscillation parameters

4.3 Analysis of Vibration Measurement

Measurement of vibrations refers to expressing of amplitude of sinusoidal signal so the following expressions are used for measurement of the overall vibrations:

– measurement of overall vibrations,
– measurement of partial vibrations,
– measurement of relative vibrations,
– measurement of absolute vibrations.

4.3.1 Measurement of Overall Vibrations

The overall vibrations represent total vibration energy measured within specific frequency range (mostly 0–1000 Hz). Frequency range within which the measurement of overall vibrations is realized depends on type of the used monitoring device.

By measurement of the overall vibrations of the machine or of its parts and by comparing the value with its normal level (according to standards), the information on objective technical condition of the machine shall be obtained and the reason of higher values shall be detected. Overall vibrations include summary effect of vibrations of its structural parts which are caused by machine operation and by impulses produced by the machine environment (environment off the machine), and they are examined by broadband frequency analysis, for instance, within the frequency range from 10 Hz up to 10 kHz. Overall vibrations offer overall yet orientation and rather inaccurate information on vibrations according to which it is impossible to identify vibrating structural part or component.

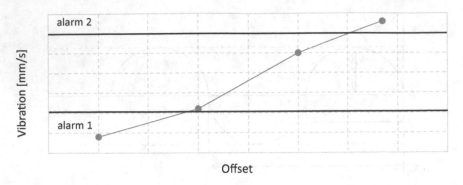

Fig. 4.2 Trend graph

Probably the most suitable and the most reliable method of assessment of intensity of vibrations is comparison of last overall measured values of vibrations with the ones measured previously which allows observing development of vibrations during specific time period. The analysis of such comparison made between current values with the previous ones is easier when the values are plotted in the trend graph (Fig. 4.2).

4.3.2   Measurement of Partial Vibrations

Partial vibrations offer accurate information on vibrations of particular structural part or component of the machine. They are caused by vibrations of the individual structural parts of an object and are examined by narrow-band frequency analysis within the frequency range, for instance, from 8 to 8.5 kHz. Complete assessment of the machine requires measurement of overall and partial vibrations—i.e. broad- and narrow-band frequency analysis should be applied.

4.3.3   Measurement of Relative Vibrations

Sensors of relative vibrations are positioned in couples so that angle of 90° is formed. The alignment allows observation of time development of vibrations as well as position of rotor or trajectory of movement of the rotor centre in the plane with the sensors. The measurement principle of relative vibrations and positioning of the sensors are shown in Fig. 4.3.

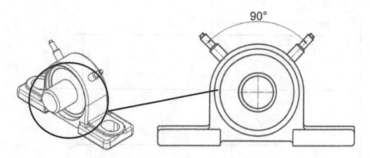

Fig. 4.3 Instance of positioning of sensors in measurement of relative vibrations

4.3.4 Measurement of Absolute Vibrations

Measurement of absolute vibrations as vibrations of machine or of machine parts with regard to stable basis can be realized by means of sensors of position, of speed or of acceleration. The given quantities can be reciprocally transferred by their derivation or integration according to time.

4.4 Analysis in Time and Frequency Sphere

The chapter contains detailed description of analysis of the values measured in time and frequency sphere.

4.4.1 Analysis in Time Domain

The analysis in time domain is based on assessment of parameters of time development of signals determining the quantities such as derivation, speed and acceleration. The methods of calculation in the time domain usually manipulate with time development of the signal or of its envelope and make an effort to describe the properties of the signal which are derived from integral quantities such as mean and effective value of a signal, energy or coefficient of amplitude of oscillation. In case of prevailing of random signal part—random vibrations—it is possible to apply for analysis the selected statistical calculations of descriptors such as standard deviation, coefficient of sharpness, coefficient of angularity or coefficient of amplitude of oscillation. Analysis of vibrodiagnostic signal in the time domain (Fig. 4.4) is suitable for transitional phenomena, and thus, a dominant vibration source exists which prevents loss of information in noise of signal. Advantage of time development is much information on individual sources. The frequency analysis is used for time signal analysis

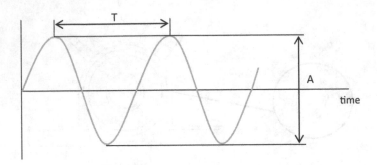

Fig. 4.4 Plotting of time domain

in case of which the frequency parts contain much information on condition of the equipment. The simplest periodical signal in time distribution can be depicted by a sinusoid which determines time development of the observed quantity. Time signals are further on processed by means of fast Fourier transform. In the time domain, the signals can be divided to assessment of time development and assessment of overall oscillation. The overall oscillation is connected with all frequencies of oscillation in the respective measured spot. The measured value of overall oscillation is compared with the previous measurement in case of which the equipment worked without defect. Further comparison is performed with the set critical values.

4.4.2 *Analysis in Frequency Domain*

The most significant tools for detection of periodic phenomena in vibration signal include frequency analysis which is applied as the main tool for locating and trending of sources. The methods of calculation in the frequency domain are usually based on discreet Fourier transform. One of the main drawbacks of the extracted indicators is impossibility to record the behaviour of non-stationary signals. Frequency analysis performs disintegration of original time development into the individual harmonic components. The total of all analysed harmonic components leads to formation of original time development. Just particular frequency components are selected from the signal spectrum, which carry required discriminant information (Fig. 4.5). Other approach of calculation of indicators in frequency domain is application of filters and calculation of integral descriptions of signals in case of their outputs. The advantage of signal assessment in the frequency domain rests in the fact that the individual phenomena are separated from each other. In the assessment of signal in time domain, the mixture of the individual phenomena can be observed.

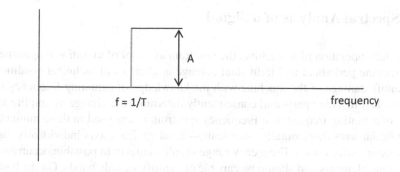

Fig. 4.5 Plotting of frequency domain

Plotting of frequency analysis is represented by frequency spectrum offering detailed information on signal sources which cannot be obtained from time signal. If the quantity is assessed in frequency domain, so with low frequencies it is appropriate to assess the amplitude and in case of high-frequency components of oscillation the assessment should be focused on effective value of acceleration.

4.5 Possibilities of Oscillation Elimination

Currently, there exist several methods and means serving for oscillation elimination in the cutting process. In general, the methods can be categorized into two phases. During the first phase, the material properties of the workpiece and of the tool are taken into consideration as well as all properties and characteristics which must be expressed prior to performance of the second phase to assure planning of the measurement setting.

The second phase refers to monitoring of the oscillation indicator during particular cutting process, for instance, on the basis of sound spectrum in machining, surface roughness, simulation, by using of the accelerometer and of the impulse hammer.

Possibilities of elimination of induced vibrations:

– selection of suitable foundation of the equipment. A flexible element is placed under the equipment which prevents transfer of oscillation from the environment,
– suitable specification of working conditions yet such parameters must be selected so that the vibration free process is assured in order to avoid resonance domain,
– increase of system stiffness.

Possibilities of elimination of self-actuated vibrations:

– change of cutting parameters,
– use of hydraulic and mechanical dampers of diverse structure.

4.6   Spectral Analysis of a Signal

In periodical operation of a machine, the spectrum as a tool of vibration diagnostics can determine periodical and individual damage or changes of technical condition and identify damage of the machine with the knowledge of actuating frequency of the individual machine parts and consequently determine the change of amplitudes in case of actuating frequencies. Frequency spectrum is analysed in three mutually perpendicular axes (horizontally—vertically—axially). Each axis individually carries important information. Frequency range should conform to possible occurrence of harmonic elements and should be capable of identifying side bands. On the basis of frequency spectrum, the following can be specified:

– *basic actuating frequencies*
   conform to failure frequencies determined by calculation from the structural parameters. They represent the function of rotor frequencies of the shaft which rotates at constant frequency. They occur in low-frequency domain of the spectrum and are used for identification of the damage source.
– *harmonic frequencies*
   representing an integer multiple of basic frequency. They are the consequence of deviations of time development from the shape of function sin (*t*). The rectangular time development equals to considerable amount of harmonic frequencies. Their magnitude in relation to basic frequency represents essential sign of damage.
– *subharmonic frequencies*—forming integer ratio of basic frequency. They are the consequence of deviations of time development from the shape of function sin (*t*). Their magnitude in relation to basic frequency represents essential sign of damage.
– *interharmonic frequencies*
   representing no integer multiple or ratio of basic frequency. They can be caused by a signal produced by an unknown source. They occur especially in advanced phase of damage.
– *frequencies occurring in side bands*
   spreading around basic or harmonic part. They occur in a constant distance to both sides (in direction to higher and lower frequencies). Their amplitude decreases with the distance. Their amount and magnitude refer to a phase of damage. Presence of side bands stems from the amplitude modulation of the signal (periodical changes of amplitude in time). The change of amplitude is mostly connected with rotor frequency. The change of amplitude with rotor frequency is shown in Fig. 4.6.
– *continual bands*
   represent consequence of non-stationary signals (random signals). They can be caused by frequency modulation. They are mostly formed by friction, by oil whirl or by flowing of liquids.

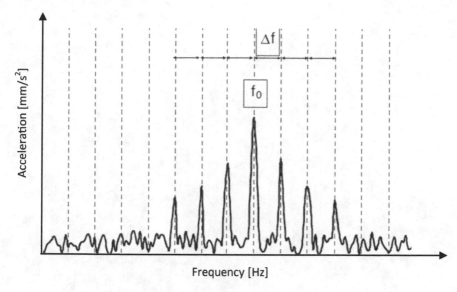

Fig. 4.6 Change of amplitude with rotor frequency

Chapter 5
Methodology of Measurement and Examining Methods

The methodology of measurement contains the basic information on performance of measurement of magnitude of acceleration of technological head vibrations of the selected technical system in the time and frequency domain without use of abrasive fraction and with the application of narrow fraction of abrasive. The chapter is divided into several basic parts:

– conditions of performance of the individual experiments,
– preparation procedure of the individual measurement,
– description of measurement spots,
– description of procedure of processing and assessment of the measured values of amplitudes of vibration acceleration.

5.1 Conditions of Performance of the Individual Experiments

All measurements of magnitude of amplitude of vibration acceleration in case of technological head were performed once the following conditions had been met:

– assurance of appropriate operating conditions of water jet technology,
– correct preparation of mass flows of abrasive for the selected abrasive types in performance of measurements by means of the storage tank and dosing device of abrasive material,
– correct preparation of narrow fraction of abrasive by means of sieve analysis,
– sensor accurately positioned on the technological head and designed for measurement of magnitude of amplitude of vibration acceleration,
– check of connection of the modular system NI 9233 by USB 2.0 to a laptop,

- adjustment of the CNC control system prior to machining according to proposed constant and changing input material and technological factors in the individual measurement files,
- sufficient measurement of time development of amplitude of acceleration of the technological head vibrations.

5.2   Process of Preparation of the Individual Measurements

Preparation of the individual measurements proceeded as follows:

- preparation of plan of experiments and of their conditions,
- preparation of mass flows of abrasive for the selected abrasive types with their grain composition which were used on measurement by means of storage tank and dosing device of abrasive material and without application of narrow fraction of abrasive,
- preparation of mass flows of abrasive for the selected types of abrasive with their grain composition used in measurement by means of weighing without and with the application of narrow fraction of the abrasive fraction,
- formation of narrow fraction of abrasive by means of sieve analysis from the mesh size of 300 to the mesh size of 200 and preparation of the examined mass flows of abrasive by means of dosing device and storage tank,
- positioning of the storage tank and of the dosing device of abrasive material,
- positioning of the sensor on the technological head which was designed for measurement of magnitude of amplitude acceleration of vibrations,
- fixation and description of the machining material to the work table of the water jet,
- connection of modular system NI 9233 by means of USB 2.0 to the laptop,
- adjustment of the work table X-Y WJ 1020-Z-EKO along with the CNC control system.

5.2.1   Preparation of Plan of Experiments

Measurement of magnitude of amplitudes of vibration acceleration in case of technological head in material machining by the water jet technology was performed with the pre-set input technological factors. The parameters were changed, i.e. abrasive mass flow, abrasive type and narrow fractions of abrasive grain by means of sieve analysis. During measurement of magnitude of amplitude of vibration acceleration, the steel HARDOX 500 with thickness of 10 mm was machined. The plan of experiments and of their conditions is schematically shown in Table 5.1.

Table 5.1 Conditions of measurement

Measurement stage	Material	Pressure (MPA)	Shift speed (mm/min)	Type of abrasive	Grain composition of abrasive	Abrasive mass flow (g/min)	Monitored quantities
1st measurement	HARDOX 500 (10 mm)	380	100	Australian garnet	MESH 80 (meshes of 300)	200 400 600	Magnitude of vibrations
2nd measurement	HARDOX 500 (10 mm)	380	100	Ukrainian garnet	MESH 80 (meshes of 300)	200 400 600	Magnitude of vibrations
3rd measurement	HARDOX 500 (10 mm)	380	100	Australian garnet	MESH 80 (meshes of 200)	200 400 600	Magnitude of vibrations
4th measurement	HARDOX 500 (10 mm)	380	100	Ukrainian garnet	MESH 80 (meshes of 200)	200 400 600	Magnitude of vibrations

5.2.2 Preparation of Mass Flows of Abrasive by Means of Storage Tank and Dosing Device

By means of storage tank and dosing device of abrasive material (Fig. 5.1) and after the interval of one minute, the scale was rotated by a micrometric screw from 0 up to 9.5 rotations at value of 0.5 in case of which the measurement of grams of the selected abrasive passing through the abrasive head was performed. When the time interval of one minute has elapsed, the amount of the abrasive having sunk through the opening of the respective angular rotation was weighed in a beaker and the obtained values were processed in the table program of Microsoft Excel and the curve was formed. By means of polynomial order, the trend curve was transferred that led to polynomial equation which was used for calculation of angular rotation of the micrometric screw with the scale for the individual pre-planned mass flows of abrasive in the individual measurement files.

Preparation of the individual pre-selected mass flows of abrasive for the individual measurements of magnitude of amplitude of acceleration of the technological head vibrations by means of storage tank and dosing device is as follows:

– preparation of mass flows for the first measurement,
– preparation of mass flows for the second measurement,
– preparation of mass flows for the third measurement,
– preparation of mass flows for the fourth measurement.

The preparation of mass flows of abrasive for the aforementioned measurements is presented in the same structure, i.e. table of weighed amounts after weighing and rotating of micrometric screw, graphical dependence of angular rotation of micrometric screw on the amount of the abrasive which sunk through the opening of dosing device at time interval of 1 min and the table of the calculated values of angular rotation of the micrometric screw after the trend curve was transferred.

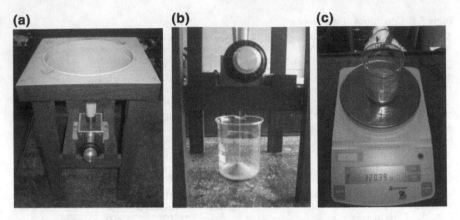

Fig. 5.1 Preparation of mass flows of abrasive (**a** storage tank and dosing device of abrasive, **b** abrasive sinking through the opening of dosing device, **c** weighing of abrasive)

5.2.2.1   Preparation of Mass Flows of Abrasive for the First Measurement

Table 5.2 presents the measured values after weighing and angular rotating of the micrometric screw for the first measurement, i.e. Australian garnet with grain composition of MESH 80 (with meshes of 300). Figure 5.2 shows the graph of dependence of angular rotation of the micrometric screw with the scale on the amount of the abrasive which sunk through the opening of the dosing device at time interval of 1 min. Table 5.3 presents the values of angular rotation of the micrometric screw with the scale which were calculated by employing the equation of the fifth polynomial for the individual planned mass flows of abrasive in the first measurement file.

Table 5.2 Weighed abrasive Australian garnet MESH 80 (meshes of 300) for the first measurement

Weight of abrasive without beaker [g]	Angular rotation of the micrometric screw	Weight of abrasive without beaker [g]	Angular rotation of the micrometric screw
25.26	3	300.44	6.5
44.28	3.5	361.7	7
70.75	4	450.18	7.5
105.17	4.5	549.99	8
140.97	5	647.27	8.5
184.09	5.5	822.77	9
235.15	6	959.5	9.5

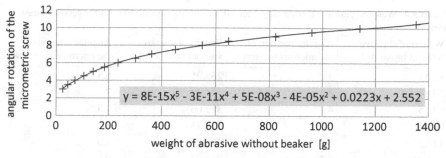

+ Australian garnet with grain composition of MESH 80
— Polynomial - Australian garnet with grain composition of MESH 80

Fig. 5.2 Envelope of the assessed values of angular rotation of the micrometric screw with the scale in dependence on the amount of the abrasive Australian garnet_MESH 80 (meshes of 300)

Table 5.3 Calculated values of angular rotation of the micrometric screw with the scale for the individual proposed flows of abrasive in case of the first measurement

Mass flows of abrasive [g/min]	Angular rotation of the micrometric screw
200	5.8
400	7.3
600	8.3

5.2.2.2 Preparation of Mass Flows for the Second Measurement

Table 5.4 presents the measured values after weighing and angular rotating of the micrometric screw for the second measurement, i.e. Ukrainian garnet with grain composition of MESH 80 (with meshes of 300). Figure 5.3 shows the graph of dependence of angular rotation of the micrometric screw with the scale on the amount of the abrasive which sunk through the opening of the dosing device at time interval of 1 min. Table 5.5 presents the values of angular rotation of the micrometric screw with the scale which were calculated by employing the equation of the fifth polynomial for the individual planned mass flows of abrasive in the second measurement file.

5.2.2.3 Preparation of Mass Flows of Abrasive for the Third Measurement

Table 5.6 presents the measured values after weighing and angular rotating of the micrometric screw for the first measurement, i.e. Australian garnet with grain composition of MESH 80 (with meshes of 200). Figure 5.4 shows the graph of dependence of angular rotation of the micrometric screw with the scale on the amount of the abrasive which sunk through the opening of the dosing device at time interval of 1 min. Table 5.7 presents the values of angular rotation of the micrometric screw with the

Table 5.4 Weighed abrasive Ukrainian garnet MESH 80 (meshes of 300) for the second measurement

Weight of abrasive without beaker [g]	Angular rotation of the micrometric screw	Weight of abrasive without beaker [g]	Angular rotation of the micrometric screw
22.41	3	255.95	6.5
39.84	3.5	321.49	7
62.15	4	397.06	7.5
91.06	4.5	473.75	8
119.27	5	578.87	8.5
164.45	5.5	768.97	9
204.85	6	891.48	9.5

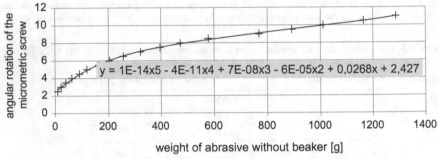

$$y = 1E\text{-}14x5 - 4E\text{-}11x4 + 7E\text{-}08x3 - 6E\text{-}05x2 + 0,0268x + 2,427$$

weight of abrasive without beaker [g]

+ Ukrainian garnet with grain composition of MESH 80
— Polynomial - Ukrainian garnet with grain composition of MESH 80

Fig. 5.3 Envelope of the assessed values of angular rotation of the micrometric screw with the scale in dependence on the amount of the abrasive Ukrainian garnet_MESH 80 (meshes of 300)

Table 5.5 Calculated values of angular rotation of the micrometric screw with the scale for the individual proposed flows of abrasive in case of the second measurement

Mass flows of abrasive [g/min]	Angular rotation of the micrometric screw
200	5.9
400	7.5
600	8.5

scale which were calculated by employing the equation of the fourth polynomial for the individual planned mass flows of abrasive in the third measurement file.

5.2.2.4   Preparation of Mass Flows of Abrasive for the Fourth Measurement

Table 5.8 presents the measured values after weighing and angular rotating of the micrometric screw for the second measurement, i.e. Ukrainian garnet with grain

Table 5.6 Weighed abrasive Australian garnet MESH 80 (meshes of 200) for the third measurement

Weight of abrasive without beaker [g]	Angular rotation of the micrometric screw	Weight of abrasive without beaker [g]	Angular rotation of the micrometric screw
25.02	3	301.11	6.5
47.67	3.5	368.51	7
72.21	4	453.61	7.5
103.11	4.5	574.42	8
141.38	5	731.62	8.5
187.34	5.5	840.67	9
239.94	6	949.62	9.5

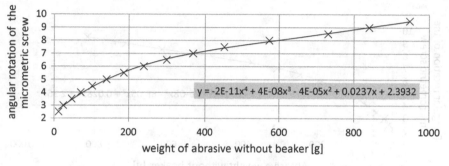

$$y = -2E\text{-}11x^4 + 4E\text{-}08x^3 - 4E\text{-}05x^2 + 0.0237x + 2.3932$$

×   Australian garnet after sieving (meshes of 200)

———   Polynomial - Australian garnet after sieving (meshes of 200)

Fig. 5.4 Envelope of the assessed values of angular rotation of the micrometric screw with the scale in dependence on the amount of the abrasive Australian garnet_MESH 80 (meshes of 200)

Table 5.7 Calculated values of angular rotation of the micrometric screw with the scale for the individual proposed flows of abrasive in case of the third measurement

Mass flows of abrasive [g/min]	Angular rotation of the micrometric screw
200	5.8
400	7.3
600	8.0

composition of MESH 80 (with meshes of 200). Figure 5.5 shows the graph of dependence of angular rotation of the micrometric screw with the scale on the amount of the abrasive which sunk through the opening of the dosing device at time interval of 1 min. Table 5.9 presents the values of angular rotation of the micrometric screw with the scale which were calculated by employing the equation of the fourth polynomial for the individual planned mass flows of abrasive in the fourth measurement file.

Table 5.8 Weighed abrasive Ukrainian garnet MESH 80 (meshes of 200) for the fourth measurement

Weight of abrasive without beaker [g]	Angular rotation of the micrometric screw	Weight of abrasive without beaker [g]	Angular rotation of the micrometric screw
21.07	3	248.51	6.5
41.73	3.5	309.6	7
65.28	4	378.71	7.5
85.3	4.5	488.1	8
117.71	5	619.87	8.5
157.26	5.5	716.24	9
198.23	6	819.8	9.5

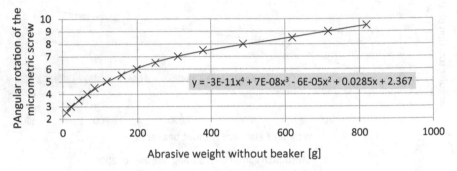

$$y = -3E\text{-}11x^4 + 7E\text{-}08x^3 - 6E\text{-}05x^2 + 0.0285x + 2.367$$

Abrasive weight without beaker [g]

× Ukrainian garnet after sieving (meshes of 200)
— Polynomial - Ukrainiangarnet after sieving(meshes of 200)

Fig. 5.5 Envelope of the assessed values of angular rotation of the micrometric screw with the scale in dependence on the amount of the abrasive Ukrainian garnet_MESH 80 (meshes of 200)

Table 5.9 Calculated values of angular rotation of the micrometric screw with the scale for the individual proposed flows of abrasive in case of the fourth measurement

Mass flows of abrasive [g/min]	Angular rotation of the micrometric screw
200	6.2
400	7.5
600	8.5

5.2.3 Preparation of Mass Flows of Abrasive by Weighing

After calculation of the angular rotation of the micrometric screw on the storage tank and on the dosing device of abrasive material by transferring of the trend curve, the examined mass flows of the individual abrasive types (meshes of 300 and of 200) were weighed and prepared into plastic bottles (Fig. 5.6) used in cutting of steel HARDOX 500 for the individual measurement files.

Fig. 5.6 Prepared mass flows of abrasive (Australian garnet_ MESH 80_ meshes of 300)

5.2.4 Formation of Narrow Fraction of Abrasive by Means of Sieve Analysis

By using the device Preciselect (Fig. 5.7) and by applying the sieve analysis, the narrow fraction of the selected types of abrasive was formed—Australian and Ukrainian garnet with grain composition of MESH 80. The sieves with the mesh size of 300 and of 200 were used for sieving (Fig. 5.8). Once the time interval elapsed, the amount of the sunken abrasive through the sieve with mesh size of 300 to the sieve with mesh size of 200 was prepared for the individual examined weight flows of abrasive by employing the similar process which involved angular rotation of the micrometric screw with the scale. The micrometric screw was used to rotate the scale from 0 up to 9.5 rotations at value of 0.5 in case of which the measurement of grams of the selected abrasive passing through the abrasive head was performed (see Sect. 5.2.2). After weighing, the obtained values were processed in the table program of Microsoft Excel and the curve was formed. By means of polynomial order, the trend curve was transferred that led to equation of the fourth polynomial on the basis of which the angular rotation of the micrometric screw on the dosing device was calculated (see Sect. 5.2.2).

5.2.5 Positioning of the Abrasive Material Storage Tank and Dosing Device

The storage tank and the dosing device of abrasive material was placed onto the work table WJ 1020-1Z-EKO as it is shown in Fig. 5.9. A funnel shape of the dosing device assures continuous discharge of the dosing device during need of mass flow of the abrasive. The micrometric screw with the scale is used for control of abrasive mass flow. The micrometric screw controls the slight opening of the gap for the abrasive flow in the bottom part of the dosing device. The abrasive in the tank is pushed by the self-weight, and consequently, it is drawn into the mixing chamber by means of a conduit.

Fig. 5.7 Presiselect device
for abrasive sieving

Fig. 5.8 Sieve with mesh
size of 300

Fig. 5.9 Positioning of the
storage tank and of the
dosing device of abrasive

5.2.6 Positioning of A Sensor for Measurement of Magnitude of Vibration Acceleration Amplitude

To monitor the magnitude of amplitude of acceleration of the technological head vibrations a miniature piezoelectric accelerometer 4507-B-004 produced by the company of Brüel and Kjær was used. The accelerometer was attached to the technological head of water jet by a quick bonding adhesive so that its axis is in congruence with the axis of vibrations in the direction of abrasive water jet (Fig. 5.10). The main technical parameters of the piezoelectric accelerometer are shown in Table 5.10.

5.2.7 Fixation of the Machining Material to A Work Table of the Water Jet

The abrasion-resistant steel of the HARDOX 500 type was used for machining. The steel was fixed to the work table WJ 1020-1Z-EKO by the binding posts shown in Fig. 5.11. The overview of basic properties of the steel is shown in Table 5.11.

Fig. 5.10 A close-up view at attachment of the miniature sensor of vibration acceleration onto the technological head

Table 5.10 Technical parameter of piezoelectric accelerometer

Frequency range	Temperature range	Nominal input	Resonant frequency	Weight
1 Hz–10 kHz	−55 °C up to +175 °C	100 mV/g	30 kHz	5.7 g

Fig. 5.11 Method of fixation of cutting material HARDOX 500 onto the work table

Table 5.11 Mechanical properties of steel HARDOX 500

Overview of properties of steel HARDOX 500							
Chemical composition in % of weight							
C	Si max.	Mn	P max.	S max.	Cr max.	Mo max.	Ni max.
0.27	0.70	1.60	0.025	0.010	1.00	0.25	0.25
Hardness							
After cold moulding max. 470 HB				After hot moulding max. 530 HB			
Mechanical properties							
Re (yield point) min. 1300 MPa				Rm (ultimate strength) min. 1550 MPa			
Carbon equivalent							
CEV = 0.62				CET = 0.41			

5.2.8 Connection of the Modular System NI 9233 by USB 2.0

The measured data are monitored by the accelerometer connected to the AD converter (AI ±5 V IEPE, sampling 25 kSps). Data record is produced and saved in the PC. A hardware core of the system is composed on the basis of the modular system CompactDAQ NI 9233 in case of which the analogue signal was transformed to a digital record. Communication with the computer is assured by the USB 2.0 interface at speed of 480 Mbit/s. The measurement card module NI 9233 is shown in Fig. 5.12, and its main technical parameters are included in Table 5.12.

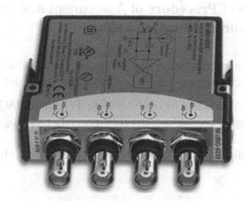

Fig. 5.12 Measurement card module NI 9233

Table 5.12 Technical parameters of measurement card module NI 9233

Technical parameters of measurement card NI 9233	
Number of channels	4
Resolution	24 bits
Range	±5 V
Actuating current	2 mA
Maximal band width	21 kHz
i/o connector	BNC connectors
Operating temperature	−40 °C + 70 °C

5.3 Description of a Measurement Spot

The measurements were performed at research and experimental workplace in a laboratory of liquid jet, Institute of Physics, Faculty of Mining and Geology, University of Mining and Metallurgy—Technical University of Ostrava. The instrumentation of the laboratory of liquid jet is equipped by the following:

- triaxial sensor of cutting forces of the abrasive water jet of own production fitted with metal and semiconducting tensometers with bridge amplifier DAQP-BRIDGE-B & PC with measurement card NI PCI-6251 M series DAQ,
- PC with measurement card NI PCI-6251 M series DAQ,
- high-pressure pump PTV 19/60 on the basis of Flow HSQ 5X pump which operates up to pressure of 415 MPa with flow rate of 1.9 l.min^{-1},
- X–Y CNC table *WJ1020-1Z-EKO* with accessories,
- pressure sensor *Kistler 4067A5000A2* with amplifier *4618A2,*
- force sensor Kistler 9301B with amplifier 5039A112,
- diamond technological head with ϕ of water nozzle amounting to 0.25 mm and ϕ of rectifier tube amounts to 1.2 mm.

5.4   Description of Procedure of Assessment and Processing of the Measured Vibration Values

Procedure of assessment and processing of the measured values of amplitude of acceleration of the technological head vibrations was realized as follows:

– processing of vibration signal by modular system CompactDAQ,
– processing of a digital record by LabVIEW program (subprogram of SignalExpress),
– processing of the digital record by table editor Microsoft Excel.

5.4.1   Processing of Vibration Signal by A Modular System

The following tools were used for the processing of vibration signal:

– miniature piezoelectric accelerometer 4507-B-004 produced by the company of Brüel and Kjær was used,
– measurement card CompactDAQ NI 9233 by the company of National Instruments,
– laptop Lenovo Z560 + power-supply units.

Modular system CompactDAQ NI 9233 is a portable and resistant device designed for processing and adjustment of a signal produced by the sensors by the company of National Instruments in which the analogue signal was transformed to a digital record. Once the setting up of the input of the accelerometer in the measurement card as well as of its specification in the application of SignalExpress was carried out, the analogue record from the accelerometer output was recalibrated during material machining (HARDOX 500) by means of a functional dependence to the amplitude of acceleration of the technological head vibrations (Fig. 5.13).

The time record intended for further processing of vibrations of the technological head was used for assessment of magnitude of vibration acceleration in the course of the entire period cutting period (Fig. 5.14).

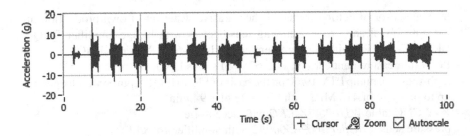

Fig. 5.13 Time record from the output of accelerometer in material cutting

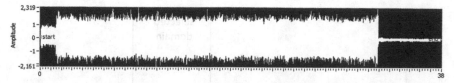

Fig. 5.14 Magnitude of amplitude of vibration acceleration in the course of entire cutting period

5.4.2   Processing of Digital Signal by the Program of SignalExpress

The software of SignalExpress was used for assessment of the measured values of the individual measurements. SignalExpress is a working environment of LabVIEW by means of which the data collection is performed from the used measurement card. The software is typical for its simple usage and possibility to analyse, to process, to record and to report the measurements.

The digital record from the analogue output of the accelerometer in machining of material HARDOX 500 was recalibrated by the functional dependence to the amplitude of acceleration of the technological head vibrations. On the basis of measurement development, the area of material penetration and of the cutting was specified (Fig. 5.15).

According to the time record intended for further processing of technological head vibrations the cutting area with stabilized course of 10 s was selected (Fig. 5.16) and assessed as magnitude of the amplitude of vibration acceleration within the frequency spectrum ranging from 0 up to 13 kHz as it is shown in graph in Fig. 5.17.

The selected part of the stabilized course of 10 s was by means of Fast Fourier Transform (hereinafter referred to as FFT) generated and assessed within the frequency spectrum ranging from 0 up to 13 Hz with the frequency step of 0.5 Hz (Fig. 5.18).

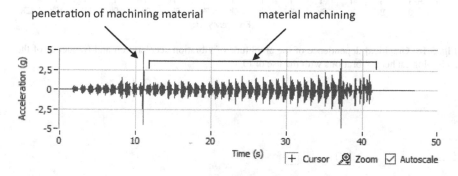

Fig. 5.15 Area of material penetration and area of material machining

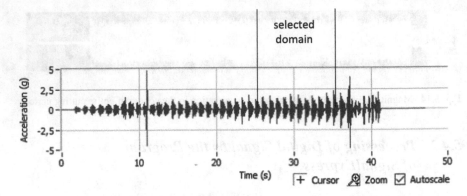

Fig. 5.16 Selected part of the stabilized course of 10 s out of the overall machining time

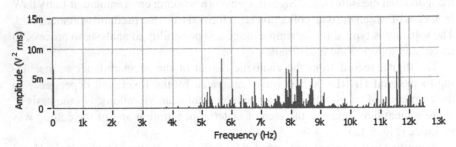

Fig. 5.17 Graphical dependence of the amplitude of vibration acceleration and frequency of the technological head vibrations without the use of FFT

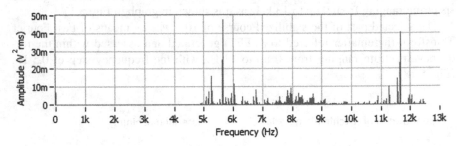

Fig. 5.18 Graphical dependence of the amplitude of vibration acceleration and frequency of the technological head vibrations with the use of FFT

Chapter 6
Assessment of Measurement of Vibration Sizes

The files of the individual measurements of the structure mentioned in Sect. 5.2.1 contain the assessment of magnitudes of amplitudes of acceleration of the technological head vibrations along with the related frequency spectra under the change of abrasive mass flow, type of abrasive and of narrow fraction of abrasive grain in case of machining of abrasion-resistant steel of K 13—HARDOX 500 type.

The developments of amplitudes of acceleration of the technological head vibrations of the water jet in the time domain and their frequency spectra are assessed for the unworn rectifier tube. The measured values are assessed within the frequency spectrum ranging from 0 Hz up to 13 Hz formed by fast Fourier transform of the selected part of the stabilized course of 10 s out of the overall time record during steel machining. They are given in the following structure:

– time records of vibrations and frequency spectra for the first measurement,
– time records of vibrations and frequency spectra for the second measurement,
– time records of vibrations and frequency spectra for the third measurement,
– time records of vibrations and frequency spectra for the fourth measurement.

6.1 Time Records of Vibrations and Frequency Spectra for the First Measurement

The chapter contains assessment of time records and of frequency spectra obtained by the software of SignalExpress on the basis of measurement of the magnitude of acceleration of the technological head vibrations in material machining with the application of the pre-specified input technological parameters—Australian garnet with grain composition of MESH 80 and with fraction of grain of the mesh of 300. The measured values are given separately for the used mass flows of the abrasive as follows:

- measured values of vibrations with abrasive mass flow of 200 g/min,
- measured values of vibrations with abrasive mass flow of 400 g/min,
- measured values of vibrations with abrasive mass flow of 600 g/min.

6.1.1 Measured Values of Vibrations with Mass Flow of 200 g/min

The assessment of basic characteristics of development of the magnitude of amplitude of vibration acceleration is shown in the graph as follows:

- time record of magnitude of acceleration of vibration amplitude (Fig. 6.1),
- frequency spectrum of amplitude of vibration acceleration (Fig. 6.2),
- envelope of frequency spectrum of amplitude of vibration acceleration (Fig. 6.3).

The time record (Fig. 6.1) referring to 0–1.2 s shows lower values of amplitude of vibration acceleration at the beginning of the section which is ascribed to measurement of vibrations without material machining. At the beginning of the section from 1.3 Hz up to 1.5 s, higher vibrations were recorded due to the first penetration of the material by the abrasive water jet. In the section referring to 1.6–10 s, the recorded development appears to be stabilized and represents the material cutting under the conditions described in Sect. 5.2.1.

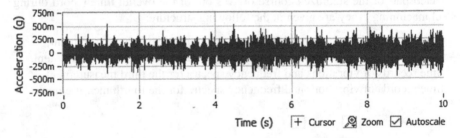

Fig. 6.1 Time record of amplitude of acceleration of vibrations for abrasive flow of 200 g/min

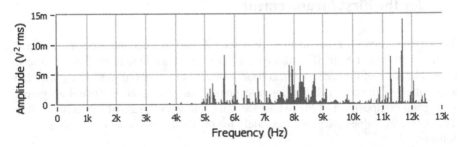

Fig. 6.2 Frequency spectrum of amplitude of acceleration of vibrations for abrasive flow of 200 g/min

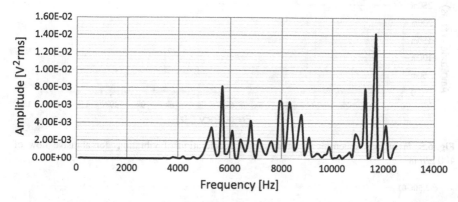

Fig. 6.3 Envelope of frequency spectrum of amplitude of acceleration of vibrations for abrasive flow of 200 g/min

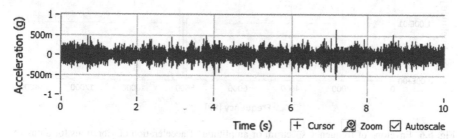

Fig. 6.4 Time record of amplitude of acceleration of vibrations for abrasive flow of 400 g/min

The measured values obtained from the frequency spectrum (Fig. 6.2) show that the first increase of the amplitude values can be observed within the frequency spectrum ranging from Z5000 to 9300 Hz. The peak value within the aforementioned range was reached with frequency of 5300 Hz 0.003489 g. In the domain from 300 Hz up to 11,000 Hz, the short increase of amplitudes was recorded to the values of up to 0.00138 g.

6.1.2   *Measured Values of Vibrations with Mass Flow of 400 g/min*

The assessment of basic characteristics of development of the magnitude of amplitude of vibration acceleration is shown in the graph, i.e. time record of magnitude of amplitude of vibration acceleration (Fig. 6.4), frequency spectrum of amplitude of vibration acceleration (Fig. 6.5) and envelope of frequency spectrum of amplitude of vibration acceleration (Fig. 6.6).

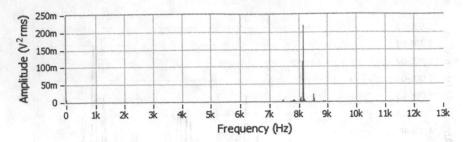

Fig. 6.5 Frequency spectrum of amplitude of acceleration of vibrations for abrasive flow of 400 g/min

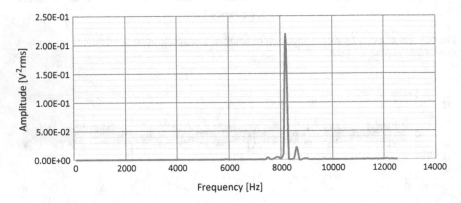

Fig. 6.6 Envelope of frequency spectrum of amplitude of acceleration of vibrations for abrasive flow of 400 g/min

The time record (Fig. 6.4) referring to 0–0.6 s shows lower values of amplitude of vibration acceleration at the beginning of the section which is ascribed to measurement of vibrations without material machining. At the beginning of the section from 0.7 up to 0.9 s, higher vibrations were recorded due to the first penetration of the material by the abrasive water jet. In the section referring to 1–10 s, the recorded development appears to be stabilized and represents the material cutting under the conditions described in Sect. 5.2.1.

The measured values obtained from the frequency spectrum (Fig. 6.5) show that during the measurement only one increased band of amplitude values was recorded within the frequency range from 8000 Hz up to 8600 Hz. The peak amplitude of 0.21855 g was observed right within the aforementioned spectrum with frequency of 8200 Hz.

6.1.3 Measured Values of Vibrations with Mass Flow of 600 g/min

The assessment of basic characteristics of development of the magnitude of amplitude of vibration acceleration is shown in the graph, i.e. time record of magnitude of amplitude of vibration acceleration (Fig. 6.7), frequency spectrum of amplitude of vibration acceleration (Fig. 6.8) and envelope of frequency spectrum of amplitude of vibration acceleration (Fig. 6.9).

The time record (Fig. 6.7) referring to 0–0.9 s shows lower values of amplitude of vibration acceleration at the beginning of the section which is ascribed to measurement of vibrations without material machining. At the beginning of the section from 1 up to 1.3 s, higher vibrations were recorded due to the first contact between material and abrasive water jet. In the section referring to 1.4–10 s, the recorded development appears to be stabilized and represents the material cutting under the conditions described in Sect. 5.2.1.

The measured values obtained from the frequency spectrum (Fig. 6.8) show that during the measurement only one increased band of amplitude values was recorded within the frequency range from 7000 Hz up to 8100 Hz. The peak amplitude of 0.307183 g was observed right within the aforementioned spectrum with frequency of 8000 Hz.

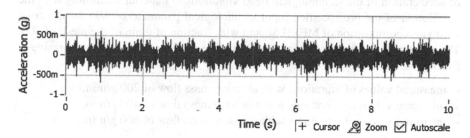

Fig. 6.7 Time record of amplitude of acceleration of vibrations for abrasive flow of 600 g/min

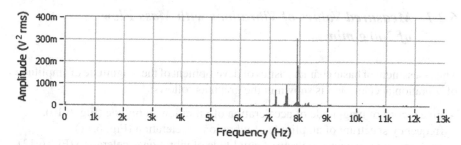

Fig. 6.8 Frequency spectrum of amplitude of acceleration of vibrations for abrasive flow of 600 g/min

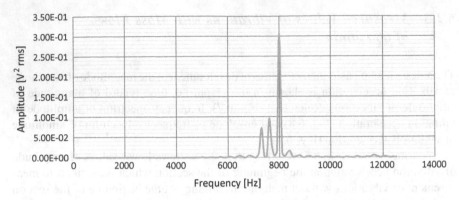

Fig. 6.9 Envelope of frequency spectrum of amplitude of acceleration of vibrations for abrasive flow of 600 g/min

6.2 Time Records of Vibrations and Frequency Spectra for the Second Measurement

The chapter contains assessment of time records and of frequency spectra obtained by the software of SignalExpress on the basis of measurement of the magnitude of acceleration of the technological head vibrations in material machining with the application of the pre-specified input technological parameters—Ukrainian garnet with grain composition of MESH 80 and with fraction of grain of the mesh of 300. The measured values are given separately for the used mass flows of the abrasive as follows:

– measured values of vibrations with abrasive mass flow of 200 g/min,
– measured values of vibrations with abrasive mass flow of 400 g/min,
– measured values of vibrations with abrasive mass flow of 600 g/min.

6.2.1 Measured Values of Vibrations with Mass Flow of 200 g/min

The assessment of basic characteristics of development of the magnitude of amplitude of vibration acceleration is shown in the graph as follows:

– time record of magnitude of acceleration of vibration amplitude (Fig. 6.10),
– frequency spectrum of amplitude of vibration acceleration (Fig. 6.11),
– envelope of frequency spectrum of amplitude of vibration acceleration (Fig. 6.12).

The time record (Fig. 6.10) referring to 0–1.2 s shows lower values of amplitude of vibration acceleration at the beginning of the section which is ascribed to measurement of vibrations without material machining. At the beginning of the section

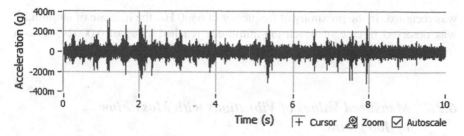

Fig. 6.10 Time record of amplitude of acceleration of vibrations for abrasive flow of 200 g/min

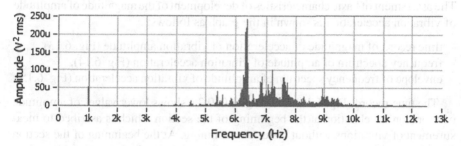

Fig. 6.11 Frequency spectrum of amplitude of acceleration of vibrations for abrasive flow of 200 g/min

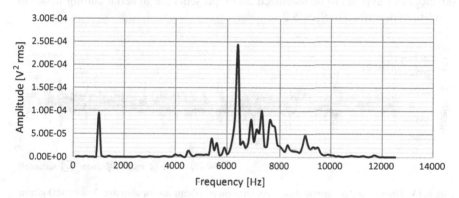

Fig. 6.12 Envelope of frequency spectrum of amplitude of acceleration of vibrations for abrasive flow of 200 g/min

from 1.3 up to 1.5 s, higher vibrations were recorded due to the first penetration of the material by the abrasive water jet. In the section referring to 1.6–10 s, the recorded development appears to be stabilized and represents the material cutting under the conditions described in Sect. 5.2.1.

The measured values obtained from the frequency spectrum (Fig. 6.11) show that the first increase of the amplitude values can be observed in the proximity of frequencies of 1000 Hz. The peak value amounted to $9.51*10^{-5}$ g. In the domain of frequency range from 5000 Hz up to 9600 Hz, the considerable increase of amplitudes was recorded. In the proximity of frequency of 6300 Hz, the increase of amplitudes was observed in case of which the amplitude reached value of 0.000243 g with frequency of 6400 Hz.

6.2.2   Measured Values of Vibrations with Mass Flow of 400 g/min

The assessment of basic characteristics of development of the magnitude of amplitude of vibration acceleration is shown in the graph as follows:

– time record of magnitude of acceleration of vibration amplitude (Fig. 6.13),
– frequency spectrum of amplitude of vibration acceleration (Fig. 6.14),
– envelope of frequency spectrum of amplitude of vibration acceleration (Fig. 6.15).

The time record (Fig. 6.13) referring to 0–0.2 s shows lower values of amplitude of vibration acceleration at the beginning of the section which is ascribed to measurement of vibrations without material machining. At the beginning of the section from 0.2 up to 0.5 s, higher vibrations were recorded due to the first penetration of the material by the abrasive water jet. In the section referring to 0.6–10 s, the recorded development appears to be stabilized and represents the material cutting under the conditions described in Sect. 5.2.1.

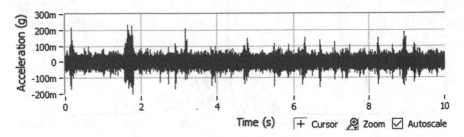

Fig. 6.13 Time record of amplitude of acceleration of vibrations for abrasive flow of 400 g/min

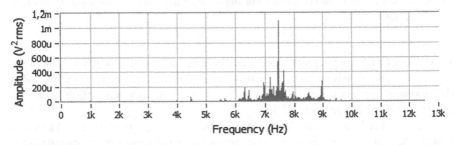

Fig. 6.14 Frequency spectrum of amplitude of acceleration of vibrations for abrasive flow of 400 g/min

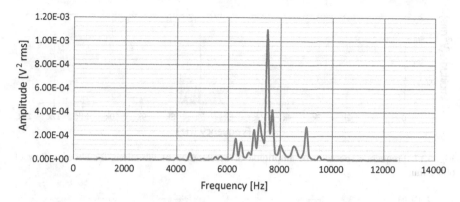

Fig. 6.15 Envelope of frequency spectrum of amplitude of acceleration of vibrations for abrasive flow of 400 g/min

The measured values obtained from the frequency spectrum (Fig. 6.14) show that the first increase of the amplitude values can be observed within the frequency range from 5700–6500 Hz. In the domain, short increase of amplitudes of up to 0.00015 could be observed. The increased band of amplitudes was visible within the range from 6800 Hz up to 8000 Hz in case of which in the proximity of frequency of 7 500 the increase of amplitudes was recorded and the amplitude value amounted to 0.001096 g. Other short increase was observed in the proximity of frequency of 9000 Hz, and the value amounted to 0.000279 g.

6.2.3 Measured Values of Vibrations with Mass Flow of 600 g/min

The assessment of basic characteristics of development of the magnitude of amplitude of vibration acceleration is shown in the graph as follows:

– time record of magnitude of acceleration of vibration amplitude (Fig. 6.16),

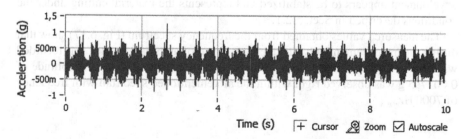

Fig. 6.16 Time record of amplitude of acceleration of vibrations for abrasive flow of 600 g/min

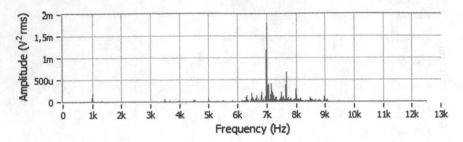

Fig. 6.17 Frequency spectrum of amplitude of acceleration of vibrations for abrasive flow of 600 g/min

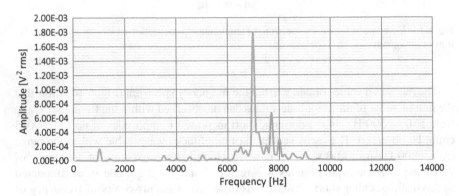

Fig. 6.18 Envelope of frequency spectrum of amplitude of acceleration of vibrations for abrasive flow of 600 g/min

- frequency spectrum of amplitude of vibration acceleration (Fig. 6.17),
- envelope of frequency spectrum of amplitude of vibration acceleration (Fig. 6.18).

The time record (Fig. 6.16) referring to 0–0.5 s shows lower values of amplitude of vibration acceleration at the beginning of the section which is ascribed to measurement of vibrations without material machining. At the beginning of the section from 0.5 up to 0.7 s, higher vibrations were recorded due to the first penetration of the material by the abrasive water jet. In the section referring to 0.7–10 s, the recorded development appears to be stabilized and represents the material cutting under the conditions described in Sect. 5.2.1.

The measured values obtained from the frequency spectrum (Fig. 6.17) show that during the measurement only one increased band of amplitude values was recorded within the frequency range from 6200 Hz up to 9000 Hz. The peak amplitude of 0.001795 g was observed right within the aforementioned spectrum with frequency of 7000 Hz.

6.3   Time Records of Vibrations and Frequency Spectra for the Third Measurement

The chapter contains assessment of time records and of frequency spectra obtained by the software of SignalExpress on the basis of measurement of the magnitude of acceleration of the technological head vibrations in material machining with the application of the pre-specified input technological parameters—Australian garnet with grain composition of MESH 80 and with fraction of grain of the mesh of 200. The measured values are given separately for the used mass flows of the abrasive as follows:

– measured values of vibrations with abrasive mass flow of 200 g/min,
– measured values of vibrations with abrasive mass flow of 400 g/min,
– measured values of vibrations with abrasive mass flow of 600 g/min.

6.3.1   Measured Values of Vibrations with Mass Flow of 200 g/min

The assessment of basic characteristics of development of the magnitude of amplitude of vibration acceleration is shown in the graph as follows:

– time record of magnitude of acceleration of vibration amplitude (Fig. 6.19),
– frequency spectrum of amplitude of vibration acceleration (Fig. 6.20),
– envelope of frequency spectrum of amplitude of vibration acceleration (Fig. 6.21).

The time record (Fig. 6.19) referring to 0–0.05 s shows lower values of amplitude of vibration acceleration at the beginning of the section which is ascribed to measurement of vibrations without material machining. At the beginning of the section from 0.05 up to 0.07 s, higher vibrations were recorded due to the first penetration of the material by the abrasive water jet. In the section referring to 0.07–10 s, the recorded development appears to be stabilized and represents the material cutting under the conditions described in Sect. 5.2.1.

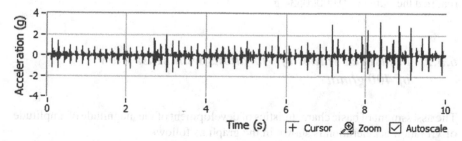

Fig. 6.19   Time record of amplitude of acceleration of vibrations for abrasive flow of 200 g/min

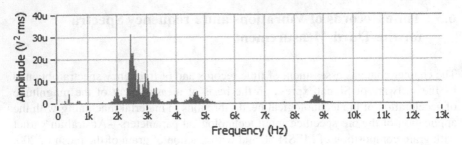

Fig. 6.20 Frequency spectrum of amplitude of acceleration of vibrations for abrasive flow of 200 g/min

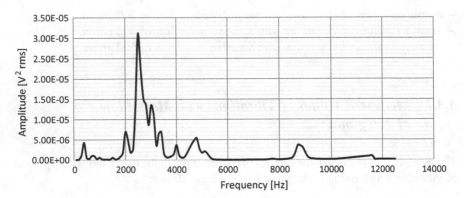

Fig. 6.21 Envelope of frequency spectrum of amplitude of acceleration of vibrations for abrasive flow of 200 g/min

The measured values obtained from the frequency spectrum (Fig. 6.20) show that the first increase of the amplitude values can be observed in the proximity of frequencies of 500 Hz. Consequently, we can observe the domain of the increased band within the range from 1900 Hz up to 4000 Hz with the peak value 0.0000310 g. In the following domain of frequency range from 4000 Hz up to 12,500 Hz, considerable increase of amplitudes was not recorded. In the proximity of frequency of 4800 Hz, the increase of amplitudes was observed in case of which the amplitude reached the value of 0.00000534 g.

6.3.2 Measured Values of Vibrations with Mass Flow of 400 g/min

The assessment of basic characteristics of development of the magnitude of amplitude of vibration acceleration is shown in the graph as follows:

– time record of magnitude of acceleration of vibration amplitude (Fig. 6.22),

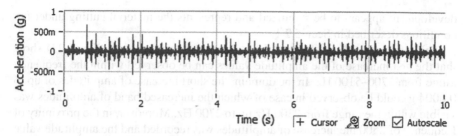

Fig. 6.22 Time record of amplitude of acceleration of vibrations for abrasive flow of 400 g/min

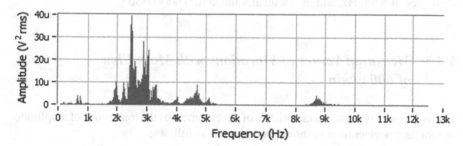

Fig. 6.23 Frequency spectrum of amplitude of acceleration of vibrations for abrasive flow of 400 g/min

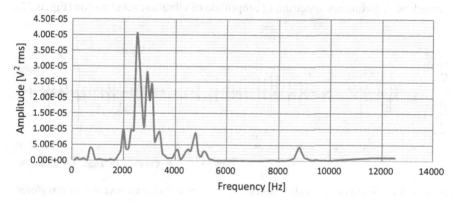

Fig. 6.24 Envelope of frequency spectrum of amplitude of acceleration of vibrations for abrasive flow of 400 g/min

– frequency spectrum of amplitude of vibration acceleration (Fig. 6.23),
– envelope of frequency spectrum of amplitude of vibration acceleration (Fig. 6.24).

The time record (Fig. 6.22) referring to 0–0.5 s shows lower values of amplitude of vibration acceleration at the beginning of the section which is ascribed to measurement of vibrations without material machining. At the beginning of the section from 0.5 up to 0.6 s, higher vibrations were recorded due to the first penetration of the material by the abrasive water jet. In the section referring to 0.7–10 s, the recorded

development appears to be stabilized and represents the material cutting under the
conditions described in Sect. 5.2.1.

The measured values obtained from the frequency spectrum (Fig. 6.23) show
that the first increase of the amplitude values can be observed within the frequency
range from 1700–5100 Hz. In the domain, the short increase of amplitudes of up to
0.0004 g could be observed in case of which the increased band of amplitudes was
visible within the range from 2300 Hz up to 2700 Hz. Moreover, in the proximity of
frequency of 2500 the increase of amplitudes was recorded and the amplitude value
amounted to 0.0000397 g. Other short increase was observed in the proximity of
frequency of 8700 Hz, and the value amounted to 0.00000436 g.

6.3.3 Measured Values of Vibrations with Mass Flow of 600 g/min

The assessment of basic characteristics of development of the magnitude of amplitude
of vibration acceleration is shown in the graph as follows:

– time record of magnitude of acceleration of vibration amplitude (Fig. 6.25),
– frequency spectrum of amplitude of vibration acceleration (Fig. 6.26),
– envelope of frequency spectrum of amplitude of vibration acceleration (Fig. 6.27).

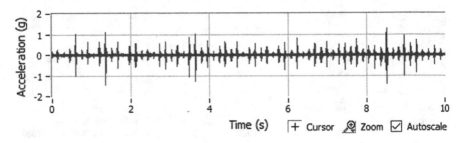

Fig. 6.25 Time record of amplitude of acceleration of vibrations for abrasive flow of 600 g/min

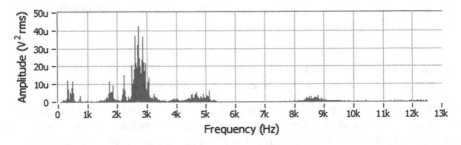

Fig. 6.26 Frequency spectrum of amplitude of acceleration of vibrations for abrasive flow of
600 g/min

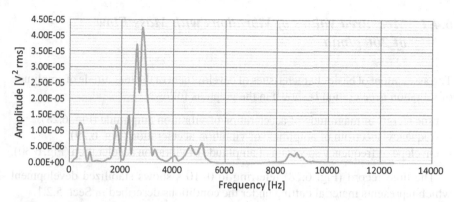

Fig. 6.27 Envelope of frequency spectrum of amplitude of acceleration of vibrations for abrasive flow of 600 g/min

The time record (Fig. 6.25) referring to 0–0.8 s shows lower values of amplitude of vibration acceleration at the beginning of the section which is ascribed to measurement of vibrations without material machining. At the beginning of the section from 0.8 up to 1 s, higher vibrations were recorded due to the first penetration of the material by the abrasive water jet. In the section referring to 1.1–10 s, the recorded development appears to be stabilized and represents the material cutting under the conditions described in Sect. 5.2.1.

The measured values obtained from the frequency spectrum (Fig. 6.26) show that during the measurement only one increased band of amplitude values was recorded within the frequency range from 100 Hz up to 5800 Hz. The peak amplitude of 0.0000421 g was observed right within the aforementioned spectrum with frequency of 2800 Hz.

6.4 Time Records of Vibrations and Frequency Spectra for the Fourth Measurement

The chapter contains assessment of time records and of frequency spectra obtained by the software of SignalExpress on the basis of measurement of the magnitude of acceleration of the technological head vibrations in material machining with the application of the pre-specified input technological parameters—Ukrainian garnet with grain composition of MESH 80 and with fraction of grain of the mesh of 200. The measured values are given separately for the used mass flows of the abrasive as follows:

– measured values of vibrations with abrasive mass flow of 200 g/min,
– measured values of vibrations with abrasive mass flow of 400 g/min,
– measured values of vibrations with abrasive mass flow of 600 g/min.

6.4.1   Measured Values of Vibrations with Mass Flow of 200 g/min

The assessment of basic characteristics of development of the magnitude of amplitude of vibration acceleration is shown in the graph as follows:

- time record of magnitude of acceleration of vibration amplitude (Fig. 6.28),
- frequency spectrum of amplitude of vibration acceleration (Fig. 6.29),
- envelope of frequency spectrum of amplitude of vibration acceleration (Fig. 6.30).

The time record (Fig. 6.28) referring to 0–10 s shows stabilized development which represents material cutting under the conditions described in Sect. 5.2.1.

The measured values obtained from the frequency spectrum (Fig. 6.29) show that the first increase of the amplitude values can be observed in the proximity of frequencies of 1600 Hz. Consequently, we can observe the domain of the increased band within the range from 3100 Hz up to 6000 Hz with the peak value 0.000125 g. In the following domain of frequency range from 6000 Hz up to 12,500 Hz, considerable increase of amplitudes was not recorded. In the proximity of frequency of 1600 Hz, the increase of amplitudes was observed in case of which the amplitude reached the value of 0.0000685 g.

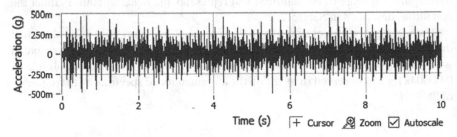

Fig. 6.28 Time record of amplitude of acceleration of vibrations for abrasive flow of 200 g/min

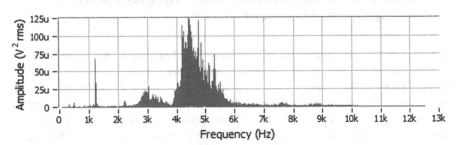

Fig. 6.29 Frequency spectrum of amplitude of acceleration of vibrations for abrasive flow of 200 g/min

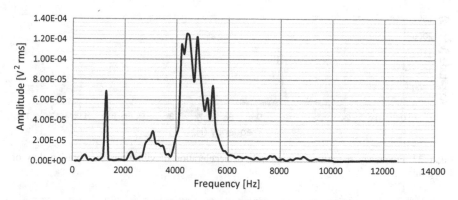

Fig. 6.30 Envelope of frequency spectrum of amplitude of acceleration of vibrations for abrasive flow of 200 g/min

6.4.2 Measured Values of Vibrations with Mass Flow of 400 g/min

The assessment of basic characteristics of development of the magnitude of amplitude of vibration acceleration is shown in the graph as follows:

– time record of magnitude of acceleration of vibration amplitude (Fig. 6.31),
– frequency spectrum of amplitude of vibration acceleration (Fig. 6.32),
– envelope of frequency spectrum of amplitude of vibration acceleration (Fig. 6.33).

The time record (Fig. 6.31) referring to 0–1 s shows lower values of amplitude of vibration acceleration at the beginning of the section which is ascribed to measurement of vibrations without material machining. At the beginning of the section from 1.1 up to 1.2 s, higher vibrations were recorded due to the first penetration of the material by the abrasive water jet. In the section referring to 1.3–10 s, the recorded development appears to be stabilized and represents the material cutting under the conditions described in Sect. 5.2.1.

The measured values obtained from the frequency spectrum (Fig. 6.32) show that the first increase of the amplitude values can be observed within frequency range

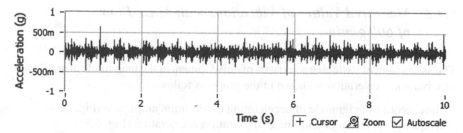

Fig. 6.31 Time record of amplitude of acceleration of vibrations for abrasive flow of 400 g/min

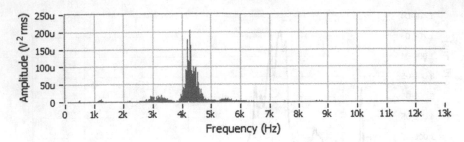

Fig. 6.32 Frequency spectrum of amplitude of acceleration of vibrations for abrasive flow of 400 g/min

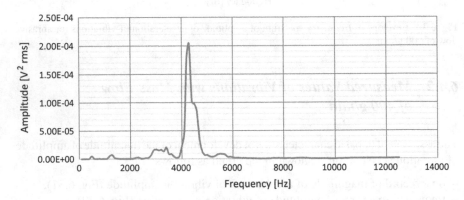

Fig. 6.33 Envelope of frequency spectrum of amplitude of acceleration of vibrations for abrasive flow of 400 g/min

from 2700 Hz up to 3700 Hz. In the domain, the short increase of amplitudes of up to 0.00000505 g could be observed. Other increased band of amplitudes was visible within the range from 3900 Hz up to 5000 Hz in case of which in the proximity of frequency of 4300 the increase of amplitudes was recorded and the amplitude value amounted to 0.000204 g. Other short increase was observed in the proximity of frequency of 5800 Hz, and the value amounted to 0.00000342 g.

6.4.3 Measured Values of Vibrations with Mass Flow of 600 g/min

The assessment of basic characteristics of development of the magnitude of amplitude of vibration acceleration is shown in the graph as follows:

– time record of magnitude of acceleration of vibration amplitude (Fig. 6.34),
– frequency spectrum of amplitude of vibration acceleration (Fig. 6.35),
– envelope of frequency spectrum of amplitude of vibration acceleration (Fig. 6.36).

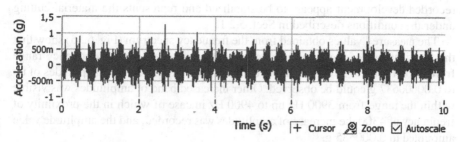

Fig. 6.34 Time record of amplitude of acceleration of vibrations for abrasive flow of 600 g/min

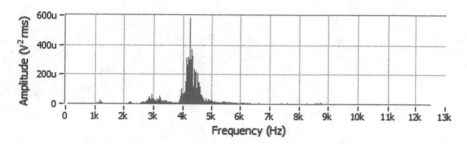

Fig. 6.35 Frequency spectrum of amplitude of acceleration of vibrations for abrasive flow of 600 g/min

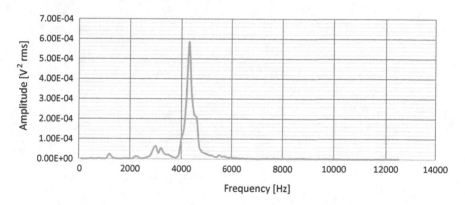

Fig. 6.36 Envelope of frequency spectrum of amplitude of acceleration of vibrations for abrasive flow of 600 g/min

The time record (Fig. 6.34) referring to 0–0.01 s shows lower values of amplitude of vibration acceleration at the beginning of the section which is ascribed to measurement of vibrations without material machining. At the beginning of the section from 0.01 up to 0.03 s, higher vibrations were recorded due to the first penetration of the material by the abrasive water jet. In the section referring to 0.04–10 s, the recorded development appears to be stabilized and represents the material cutting under the conditions described in Sect. 5.2.1.

The measured values obtained from the frequency spectrum (Fig. 6.35) show that the first increase of the amplitude values can be observed within frequency range from 2700 Hz up to 3700 Hz. In the domain, the short increase of amplitudes of up to 0.00000637 g could be observed. Other increased band of amplitudes was visible within the range from 3900 Hz up to 4900 Hz in case of which in the proximity of frequency of 4300 the increase of amplitudes was recorded, and the amplitude value amounted to 0.000585 g.

Chapter 7
Comparison of Measurements of Vibration Sizes

Comparison of the individual measurements of sizes of the acceleration amplitudes of technological head vibrations rests in formation of the envelopes in comparative graphs from the graphical dependence of frequency spectra described in the previous Chap. 6 "Assessment of Measurement of Vibration Sizes". The following structure includes the measured values:

- comparative envelopes of frequency spectra for the first and for the third measurement,
- comparative envelopes of frequency spectra for the second and for the fourth measurement.

7.1 Comparative Envelopes of Frequency Spectra for the First and for the Third Measurement

The chapter contains assessment of envelopes of frequency spectra obtained by the software of SignalExpress on the basis of measurement of the magnitude of acceleration of the technological head vibrations in material machining with the application of the pre-specified input technological parameters—Australian garnet with grain composition of MESH 80 without the use of narrow fraction of abrasive (meshes of 300) and with the use of narrow fraction of abrasive grain (meshes of 200). The measured values are given separately for the used mass flows of the abrasive as follows:

- measured values of vibrations with abrasive mass flow of 200 g/min,
- measured values of vibrations with abrasive mass flow of 400 g/min,
- measured values of vibrations with abrasive mass flow of 600 g/min.

7.1.1 Measured Values of Vibrations with Weight Flow of 200 g/min

The envelopes in comparative graph (Fig. 7.1) were formed from graphical dependences of frequency spectra by means of which compared are the measured values of vibration acceleration in material cutting with abrasive Australian garnet without fraction of abrasive with mesh of 300 and with fraction of abrasive with mesh of 200. The envelopes are formed in the entire frequency range from 0 up to 12,500 Hz as the whole band showed increased values of vibrations during cutting.

The comparative graph of envelopes of frequency spectra (Fig. 7.1) shows the values of amplitudes of vibration acceleration with narrow fraction of abrasive with mesh of 200 and without narrow fraction of grain with mesh of 300. In case of abrasive with fraction of mesh of 300 the peak value of vibration acceleration amounted to 0.014124 g with frequency of approximately 11,700 Hz. More significant value is acceleration of vibrations of 0.008172 g with frequency of approximately 5700 Hz. In case of abrasive with fraction of mesh of 200, the amplitude of acceleration of vibrations markedly dropped to the value of 0.000031 g with frequency of approximately 2200 Hz. The analysis of the comparative graph proved that in case of abrasive without fraction of abrasive with mesh of 300, the values of vibration acceleration are higher contrary to the case of abrasive with fraction of abrasive with mesh of 200.

7.1.2 Measured Values of Vibrations with Weight Flow of 400 g/min

The envelopes in comparative graph (Fig. 7.2) were formed from graphical dependences of frequency spectra by means of which compared are the measured values

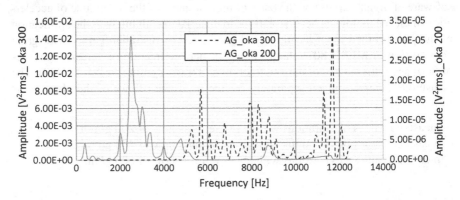

Fig. 7.1 Comparative graph of envelopes of frequency spectra for abrasive flow of 200 g/min

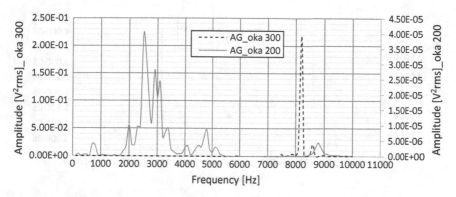

Fig. 7.2 Comparative graph of envelopes of frequency spectra for abrasive flow of 400 g/min

of vibration acceleration in material cutting with abrasive Australian garnet without fraction of abrasive with mesh of 300 and with fraction of abrasive with mesh of 200. The envelopes are formed in the entire examined frequency range from 0 up to 10,000 Hz, as in case of the band from 10,000 Hz and higher, the increased values of acceleration amplitudes of vibrations were not recorded.

The comparative graph of envelopes of frequency spectra (Fig. 7.2) shows the values of amplitudes of vibration acceleration with narrow fraction of abrasive with mesh of 200 and without narrow fraction of grain with mesh of 300. In case of abrasive with fraction of mesh of 300, the peak value of vibration acceleration amounted to 0.21855 g with frequency of approximately 8200 Hz. In case of abrasive with fraction of mesh of 200, the amplitude of acceleration of vibrations markedly dropped to the value of 0.0000397 g with frequency of approximately 2500 Hz. More significant value is vibration acceleration of 0.000028 g with frequency of approximately 2800 Hz. The analysis of the comparative graph proved that in case of abrasive without fraction of abrasive with mesh of 300, the values of vibration acceleration are higher contrary to the case of abrasive with fraction of abrasive with mesh of 200.

7.1.3 Measured Values of Vibrations with Mass Flow of 600 g/min

The envelopes in comparative graph (Fig. 7.3) were formed from graphical dependences of frequency spectra by means of which compared are the measured values of vibration acceleration in material cutting with abrasive Australian garnet without fraction of abrasive with mesh of 300 and with fraction of abrasive with mesh of 200. The envelopes are formed in the entire examined frequency range from 0 up to 10,000 Hz, as in case of the band from 10,000 Hz and higher, the increased values of acceleration amplitudes of vibrations were not recorded.

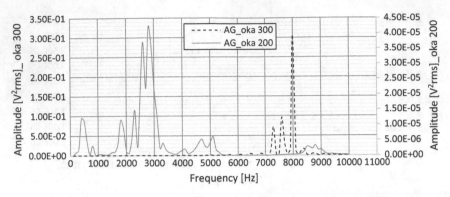

Fig. 7.3 Comparative graph of envelopes of frequency spectra for abrasive flow of 600 g/min

The comparative graph of envelopes of frequency spectra (Fig. 7.3) shows the values of amplitudes of vibration acceleration with narrow fraction of abrasive with mesh of 200 and without narrow fraction of grain with mesh of 300. In case of abrasive with fraction of mesh of 300, the peak value of vibration acceleration amounted to 0.307183 g with frequency of approximately 8000 Hz. More significant value is vibration acceleration of 0.028757 g with frequency of approximately 7700 Hz. In case of abrasive with fraction of mesh of 200, the amplitude of acceleration of vibrations markedly dropped to the value of 0.0000421 g with frequency of approximately 2800 Hz. The analysis of the comparative graph proved that in case of abrasive without fraction of abrasive with mesh of 300 the values of vibration acceleration are higher contrary to the case of abrasive with fraction of abrasive with mesh of 200.

7.2 Comparative Envelopes of Frequency Spectra for the Second and for the Fourth Measurement

The chapter contains assessment of envelopes of frequency spectra obtained by the software of SignalExpress on the basis of measurement of the magnitude of acceleration of the technological head vibrations in material machining with the application of the pre-specified input technological parameters—Ukrainian garnet with grain composition of MESH 80 without the use of narrow fraction of abrasive (meshes of 300) and with the use of narrow fraction of abrasive grain (meshes of 200). The measured values are given separately for the used mass flows of the abrasive as follows:

– measured values of vibrations with abrasive mass flow of 200 g/min,
– measured values of vibrations with abrasive mass flow of 400 g/min,
– measured values of vibrations with abrasive mass flow of 600 g/min.

7.2.1 *Measured Values of Vibrations with Weight Flow of 200 g/min*

The envelopes in comparative graph (Fig. 7.4) were formed from graphical dependences of frequency spectra by means of which compared are the measured values of vibration acceleration in material cutting with abrasive Ukrainian garnet without fraction of abrasive with mesh of 300 and with fraction of abrasive with mesh of 200. The envelopes are formed in the entire examined frequency range from 0 up to 11,000 Hz, as in case of the band from 11,000 Hz and higher, the increased values of acceleration amplitudes of vibrations were not recorded.

The comparative graph of envelopes of frequency spectra (Fig. 7.4) shows the values of amplitudes of vibration acceleration with narrow fraction of abrasive with mesh of 200 and without narrow fraction of grain with mesh of 300. In case of abrasive with fraction of mesh of 300, the peak value of vibration acceleration amounted to 0.000243 g with frequency of approximately 6400 Hz. More significant value is vibration acceleration of 0.0001 g with frequency of approximately 1000 Hz and 7200 Hz. In case of abrasive with fraction of mesh of 200, the amplitude of acceleration of vibrations markedly dropped to the value of 0.000125 g with frequency of approximately 3900 Hz. The analysis of the comparative graph proved that in case of abrasive without fraction of abrasive with mesh of 300, the values of vibration acceleration are higher contrary to the case of abrasive with fraction of abrasive with mesh of 200.

7.2.2 *Measured Values of Vibrations with Mass Flow of 400 g/min*

The envelopes in comparative graph (Fig. 7.5) were formed from graphical dependences of frequency spectra by means of which compared are the measured values of vibration acceleration in material cutting with abrasive Ukrainian garnet without

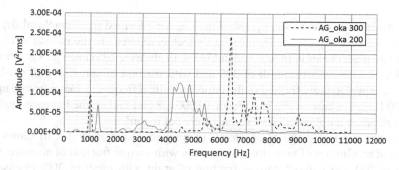

Fig. 7.4 Comparative graph of envelopes of frequency spectra for abrasive flow of 200 g/min

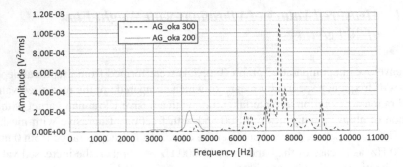

Fig. 7.5 Comparative graph of envelopes of frequency spectra for abrasive flow of 400 g/min

fraction of abrasive with mesh of 300 and with fraction of abrasive with mesh of 200. The envelopes are formed in the entire examined frequency range from 0 up to 10,000 Hz as in case of the band from 10,000 Hz and higher the increased values of acceleration amplitudes of vibrations were not recorded.

The comparative graph of envelopes of frequency spectra (Fig. 7.5) shows the values of amplitudes of vibration acceleration with narrow fraction of abrasive with mesh of 200 and without narrow fraction of grain with mesh of 300. In case of abrasive with fraction of mesh of 300, the peak value of vibration acceleration amounted to 0.001096 g with frequency of approximately 7500 Hz. More significant value is vibration acceleration of 0.000279 g with frequency of approximately 9000 Hz. In case of abrasive with fraction of mesh of 200 the amplitude of acceleration of vibrations markedly dropped to the value of 0.000204 g with frequency of approximately 4300 Hz. The analysis of the comparative graph proved that in case of abrasive without fraction of abrasive with mesh of 300 the values of vibration acceleration are higher contrary to the case of abrasive with fraction of abrasive with mesh of 200.

7.2.3 Measured Values of Vibrations with Mass Flow of 600 g/min

The envelopes in comparative graph (Fig. 7.6) were formed from graphical dependences of frequency spectra by means of which compared are the measured values of vibration acceleration in material cutting with abrasive Ukrainian garnet without fraction of abrasive with mesh of 300 and with fraction of abrasive with mesh of 200. The envelopes are formed in the entire examined frequency range from 0 up to 10,000 Hz, as in case of the band from 10,000 Hz and higher, the increased values of acceleration amplitudes of vibrations were not recorded.

The comparative graph of envelopes of frequency spectra (Fig. 7.6) shows the values of amplitudes of vibration acceleration with narrow fraction of abrasive with mesh of 200 and without narrow fraction of grain with mesh of 300. In case of

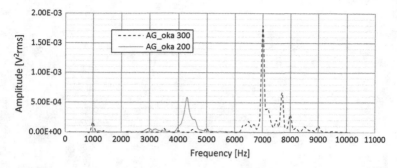

Fig. 7.6 Comparative graph of envelopes of frequency spectra for abrasive flow of 600 g/min

abrasive with fraction of mesh of 300, the peak value of vibration acceleration amounted to 0.001795 g with frequency of approximately 7000 Hz. More significant value is vibration acceleration of 0.000669 g with frequency of approximately 7700 Hz. In case of abrasive with fraction of mesh of 200, the amplitude of acceleration of vibrations markedly dropped to the value of 0.000585 g with frequency of approximately 4 300 Hz. The analysis of the comparative graph proved that in case of abrasive without fraction of abrasive with mesh of 300, the values of vibration acceleration are higher contrary to the case of abrasive with fraction of abrasive with mesh of 200.

Chapter 8
Utilization of Knowledge of the Individual Measurements of Vibration Sizes

In general, the technical condition of equipment in operation deteriorates in the course of the time. Deterioration of technical condition represents the consequence of the influence of factors with negative effect. The fundamental factors include increased stress, temperature, dust nuisance and vibrations. Monitoring of the parameters allows obtaining basic objective data inevitable for assessment of the condition of equipment or for correction of negative condition. The operative condition of equipment requires the values to range within the specified limits.

The chapter was written on the basis of results of the measured values of the size of acceleration amplitudes of the technological head vibrations of water jet which were obtained in the individual measurements. New knowledge and recommendations are focused on the elimination of the undesired influences of vibrations in cutting by means of this technology. The very design proposal for the utilization of knowledge in practice consists of the graphical dependences with maximal values of amplitudes of vibration acceleration (Figs. 8.1, 8.2, 8.3 and 8.4).

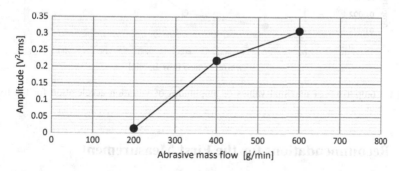

Fig. 8.1 Comparison of maximal values of amplitudes of vibration acceleration for the first measurement

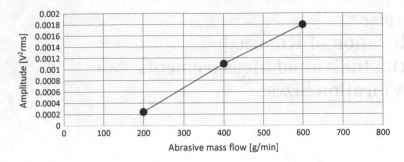

Fig. 8.2 Comparison of maximal values of amplitudes of vibration acceleration for the second measurement

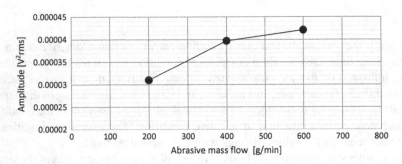

Fig. 8.3 Comparison of maximal values of amplitudes of vibration acceleration for the third measurement

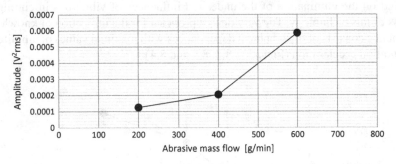

Fig. 8.4 Comparison of maximal values of amplitudes of vibration acceleration for the fourth measurement

8.1 Recommendations for the First Measurement

The recommendations are applicable for the conditions described in Sect. 5.2.1. in case of which during material cutting the three mass flows of the abrasive of Australian garnet were used with grain composition of MESH 80 without narrow fraction of abrasive grain with mesh of 300 and with shift speed of technological head amounting to 100 mm/min. HARDOX steel 500 with thickness of 10 mm is recommended, from the point of view of the elimination of the size of acceleration amplitude of technological head vibrations, for machining with abrasive mass flow of 200 g/min—with regard to comparison of maximal values of the size of vibrations with other two employed mass flows of 400 and 600 g/min. The elimination of maximal values of acceleration amplitudes of vibrations requires avoidance of frequency band within the range from 10,000 to 12,500 Hz with flow of 200 g/min. In case of abrasive flow of 400 g/min and 600 g/min, it is inevitable to avoid frequencies within the range from 7000 to 8300 Hz. Furthermore, the graphical representation of frequency spectra proved that with the increase in the value of mass flow of the abrasive of Australian garnet with mesh 300 definitely and almost regularly increases maximal value of acceleration amplitude of vibrations.

8.2 Recommendation for the Second Measurement

The recommendations are applicable for the conditions described in Sect. 5.2.1. in case of which during material cutting the three mass flows of the abrasive of Ukrainian garnet were used with grain composition of MESH 80 without narrow fraction of abrasive grain with mesh of 300 and with shift speed of technological head amounting to 100 mm/min. HARDOX steel 500 with thickness of 10 mm is recommended, from the point of view of elimination of the size of acceleration amplitude of technological head vibrations, for machining with abrasive mass flow of 200 g/min as in case of Australian garnet with mesh of 300—with regard to comparison of maximal values of the size of vibrations with other two employed mass flows of 400 and 600 g/min. The elimination of maximal values of acceleration amplitudes of vibrations requires avoidance of frequency band within the range from 4000 to 10,000 Hz with all three examined abrasive flows. Furthermore, the graphical development of frequency spectra proved that with the increase in the value of mass flow of the abrasive of Ukrainian garnet with mesh of 300 definitely and almost regularly increases maximal value of acceleration amplitude of vibrations.

8.3 Recommendations for the Third Measurement

The recommendations are applicable for the conditions described in Sect. 5.2.1. in case of which during material cutting the three mass flows of the abrasive of Australian garnet were used with grain composition of MESH 80 without narrow fraction of abrasive grain with mesh of 200 and with shift speed of technological head amounting to 100 mm/min. HARDOX steel 500 with thickness of 10 mm is recommended, from the point of view of elimination of the size of acceleration amplitude of technological head vibrations, for machining with abrasive mass flow of 200 g/min as in case of Australian garnet with mesh of 300 and with Ukrainian garnet with mesh of 200—with regard to comparison of maximal values of the size of vibrations with other two employed mass flows of 400 and 600 g/min. The elimination of maximal values of acceleration amplitudes of vibrations requires avoidance of frequency band within the range from 1500 to 6000 Hz with all three examined abrasive flows. Furthermore, the graphical development of frequency spectra proved that with the increase in the value of mass flow of the abrasive of Australian garnet with mesh of 200 definitely and almost regularly increases maximal value of acceleration amplitude of vibrations.

8.4 Recommendations for the Fourth Measurement

The recommendations are applicable for the conditions described in Sect. 5.2.1. in case of which during material cutting the three mass flows of the abrasive of Ukrainian garnet were used with grain composition of MESH 80 without narrow fraction of abrasive grain with mesh of 200 and with shift speed of technological head amounting to 100 mm/min. HARDOX steel 500 with thickness of 10 mm is recommended, from the point of view of elimination of the size of acceleration amplitude of technological head vibrations, for machining with abrasive mass flow of 200 g/min as in case of previous measurements—with regard to comparison of maximal values of the size of vibrations with other two employed mass flows of 400 and 600 g/min. The elimination of maximal values of acceleration amplitudes of vibrations requires avoidance of frequency band within the range from 4000 Hz to 6000 Hz with all three examined abrasive flows. Furthermore, the graphical development of frequency spectra proved that with the increase in the value of mass flow of the abrasive of Ukrainian garnet with mesh of 300 definitely and almost regularly increases maximal value of acceleration amplitude of vibrations.

Chapter 9
Assessment

Mechanical oscillation represents the undesired factor with negative effect upon almost any equipment with its environment. A long-term effect of oscillation upon the systems of structural complexes causes shortening of the service life and increasing of noise level. Diverse literatures and standards list limit values of two parameters of vibrations, i.e. speed and acceleration of vibration amplitudes. They specify the amount by which the value of vibrations can increase in contrast to initial value yet only a few technical devices dispose of definitions of these values. By means of configuration of theoretical knowledge and by means of performance of experiments, it was partially achieved. The formed frequency spectra and their consequent analysis define the initial state and represent a suitable basis for defining the boundary which the values of vibration sizes cannot exceed. Pursuant to description and analysis of the present state of cognition and solutions focused on operation of production system with the water jet technology, the monograph "XZ" updates current knowledge and solutions by new information in the sphere of research and application of the sieve analysis of abrasive in relation to origin and magnitude of acceleration amplitudes of technological head vibrations.

The monograph deals especially with the examination of the influence of the change of mass flow of the Australian and Ukrainian abrasive grain (200, 400 and 600 g/min) with and without narrow fraction of abrasive grain (with meshes of 300 and of 200) acting upon the formation and magnitude of acceleration amplitude of technological head vibrations. The design proposal introduced in the monograph rests in the performance and assessment of 4 main measurements with the total extent of 12 experiments. The experiments were performed with production system with the technology of abrasive water jet under laboratory conditions of Science and Research Workplace of Water Jet at Institute of Physics, Faculty of Mining and Geology, University of Mining and Metallurgy—Technical University of Ostrava. By the LabVIEW software in the application of SignalExpress, the development of acceleration amplitudes of vibrations was recorded during measurements in dependence on time by Fourier transform with frequencies of technological head transformation and consequently their processing and assessment were performed.

The achieved results proved that with the increasing abrasive mass flow the values of acceleration amplitudes of vibrations increase with as well as without the use of narrow fraction of abrasive grain. Furthermore, it was proved that with fraction of abrasive grain with mesh of 200 far lower values of acceleration amplitudes of vibrations were achieved than with fraction of abrasive grain with mesh of 300. Out of the set of experiments advisable is to use Australian garnet with grain composition of abrasive of MESH 80 with the use of narrow fraction of abrasive grain with mesh of 200. In case of frequency analysis, considerable increase of amplitudes was detected in mean frequency parts of spectra.

From the data obtained by long-term measurement, the trend of vibration acceleration amplitude increase may be determined in the future and thus using the given limits assigned to the particular natural frequencies alert the deteriorating state or localize a constructional component with increased probability of a failure. The attention paid to causes of undesired oscillation and to its elimination and to consequent avoidance represent basis of successful diagnostics of machines and equipment.

Bibliography

1. Chen FL, Siores E. The effect of cutting jet variation on striation formation in abrasive water jet cutting. Int J Mach Tools Manuf. 2001;41:1479–86.
2. Chen FL, Siores E. The effect of cutting jet variation on striation formation in abrasive water jet cutting. J Mater Process Technol 2003;35(1).
3. Crofton PSJ, Abudaka M. Theoretical analysis and preliminary experimental results for an abrasive water jet cutting head. In 5th American Water Jet Conference, Toronto, Canada August 1989;29–31.
4. Deam RT, Lemma E, Ahmed DH. Modelling of the abrasive water jet cutting process. Wear 2004; 257(9–10):877–891.
5. Fabian S, Plančár Š. Contribution to more complex diagnostics and examination of operating condition of production systems by AWJ technology. In: Operation and Diagnostics of Machines and Production Systems Operational States: Scientific Papers: Vol. 4. Lüdenscheid: RAM-Verlag, 2011, pp. 64–68. ISBN 978-3-942303-10-1.
6. Fabian S, Servátka M. The experimental stating and the analysis of mutual relevance of the technological parameters influence on the surface machined by technology AWJ roughness parameters. In: Operation and diagnostics of machines and production systems operational states: scientific papers Vol. 3. Lüdenscheid: RAM-Verlag; 2010, pp. 76–83. ISBN 978-3-942303-04-0.
7. Fabian S, Salokyová Š. Experimental investigation and analysis of the impact in abrasive mass flow changes with and without the use of sieve analysis on technological head vibrations at hydroabrasive cutting. Appl Mechan Mater Aug 2014; Zurich 616:85–92.
8. Fabian S. Salokyová Š, Olejár T. Analysis and experimental study of the technological head feed rate impact on vibrations and their frequency spectra during material cutting using AWJ technology. Nonconventional Technol Rev 2011;15(3):27–32.
9. Gokhan A. Recycling of abrasives in abrasive water jet cutting with different types of granite. Arab J Geosci 2013; 9.
10. Guo NS, Louis H, Meier G, Ohlsen J. Recycling capacity of abrasives in abrasive waterjet cutting. In: Proceedings of the 11th international conference on jet cutting technology, Amsterdam, Scotland; 1992. pp. 503–523.
11. Hace A, Jezernik K. Control system for the waterjet cutting machine. IEEE/ASME Trans Mechatron. 2004;9:627–35.
12. Hashish M. A model study of metal cutting with abrasive waterjets. ASME J Eng Mater Technol. 1984;106:88–100.
13. Hashish M. Cutting with abrasive waterjets. Mech Eng. 1984;106(3):60–9.

14. Hashish M. Optimization factors in abrasive waterjet machining. ASME J Eng Ind. 1991;113:9–37.

15. Hashish M. Pressure effects in abrasive waterjet machining. J Eng Mater Technol. 1989;111:221–8.

16. Himmelreich U, Riess W. Hydrodynamic investigations on abrasive water jet cutting tools. In: Proc. 10th Int. Jet Cutting Technol., Elsevier Applied Science, London 1990, pp. 3–22.

17. Hlaváč L. Macroscopic physical description of the high energy liquid jet interaction with material. Professors' lectures. ČVUT Praha 1/2006, p. 30. ISBN 80–01–03465–8.

18. Hlaváč L. Physical description of high energy liquid jet interaction with material. Geomechanics 1992;91:341–346 (Balkema/Rotterdam/Brookfield).

19. Hlaváč L. Physical model of jet—abrasive interaction. Geomechanics 1994;93:301–304 (Balkema/Rotterdam/Brookfield).

20. Kantha Babu M, Krishnaiah OV, Chetty. A study on recycling of abrasives in abrasive water jet machining. Wear 2003;254:763–773.

21. Kantha Babu M, Krishnaiah OV. Chetty Studies on recharging of abrasives in abrasive water jet machining. Adv Manuf Technol 2002;19:697–703.

22. Kim TJ, Sylvia JG, Posner L. Piercing and cutting of ceramics by abrasive waterjet. Miami Beach, FL: The Winter Annual Meeting of the ASME; 1985. pp. 19–24.

23. Labus TJ, Neusen KF, Alberts DG, Gores TJ. Factors influencing the particle size distribution in an abrasive water jet. Trans ASME J Eng Ind. 1991;113:402–11.

24. Lemma E, Chen L, Siores E, Wang J. Optimising the AWJ cutting process of ductile materials using nozzle oscillation technique. Int J Mach Tools Manuf. 2002;42(7):781–9.

25. Lipovszky G, Sólyomvári K, Varga G. Vibration testing of machines and their maintenance. Amsterdam: Elsevier; 1990. ISBN 0-444-98808-4.

26. Mason F. Water and sand, wet grit, abrasive waterjets, a special report. American Machinist, October 1989, pp. 84–95.

27. Momber AW, Kovacevic R. Principles of abrasive water jet machining. Berlin: Springer; 1998.

28. Momber W, Kovacevic R. Principles of abrasive water jet machining. Springer; 1997.

29. Nadeau E. Prediction and role of abrasive water jet velocity in abrasive water jet cutting. Int J Water Jet Technol. 1991;1:109–16.

30. Nouraei H, Wodoslawsky A, Papini M. Characteristics of abrasive slurry jet micro-machining: a comparison with abrasive air jet micro-machining. J Mater Process Technol. 2013;213:1711–24.

31. Osman AH, Hashish M. Visual information of the mixing process inside the AWJ cutting head. In: Proceeding of the 9th American water jet conference, Michigan. WJTA Publishing; 23–26 August 1997, pp. 189–209.

32. Panda A, Prislupčák M, Jurko J, Pandova I, Orendáč P. Vibration and experimental comparison of machining process. Key Eng Mater Zurich, October 2015; 669:179–186.

33. Raissi K, Basile G, Cornier A, Simonin O. Abrasive air water jet modulization. In: Proceedings of the 8th American water jet conference, Houston, Texas. 1995. pp. 153–170.

34. Ramulu M, Arola D. A study of kerf characteristics in abrasive water jet machining of graphite/epoxy composite. J Eng Mater Technol. 1996;118:256–65.

35. Salokyová Š. The verification of different abrasives types impact on frequency spectrum vibrations. Acad J Manuf Eng. 2013;11(1):108–13.

36. Salokyová Š. Analysis, modelling and simulation of vibration in production systems with water jet technology. Dissertation. FVT TUKE, Prešov; 2012.

37. Schwetz KA, Greim J, Sigl LS, Pontvianne PM, Ehlbeck U, Basile G, Raissi K, Slotte P. Research on design and application of industrial scale hydro-abrasive jet-cutting nozzles. In: Proceedings of the 12th international conference on jet cutting technology, Rouen, France. 1994. pp. 165–175.

38. Simpson M. Abrasive particle study in high pressure water jet cutting. Int J Water Jet Tech. 1991;1:17–28.

39. Singh PJ. Relative performance of abrasives in abrasive water jet cutting. In: Proceedings, 12th international conference on jet cutting technology, France. 1994. pp. 521–541.

40. Tazibt AF, Mech J. Prediction of abrasive particle velocity in a high-pressure water jet and effect of air on acceleration process 1996; 15(4):527–543.

41. Vasilko K, Kmec J. Material Cutting. Technology of Cutting. Datapress Prešov 2003, ISBN 80–7O99–903–9.

42. Vijay M, Wenzhou Y. Water jet cutting techniques for processing of hard rock material. Int J Surf Min. 1989;3:59–69.

43. Zeng J, Kim TJ. Machinability of engineering materials in abrasive water jet machining. Int J Water Jet Technol. 1995;2(2):103–10.

44. STN 011312 Mechanical vibration and shocks. Marks and units.

45. STN 011390 Measurement of mechanical vibrations.

46. STN ISO 5348 Mechanical vibration and shock. Mechanical mounting of accelerometers.

47. STN 01010 Terminology of technical diagnostics.

48. STN 010103—75 Calculation of reliability parameters in technology.

49. STN 011301—73 Quantities and units in science and engineering practice.

50. Workplace HGF VŠB-TUO. [cit. 15-11-2011]. Available at: http://hgf10.vsb.cz/516/vybav.html.

51. Scientific research focus of liquid jet department. [cit. 15-11-2011]. Available at: http://if.vsb.cz/Veda/paprsek.html.

Printed in the United States
By Bookmasters